Governing Complex Systems

Earth System Governance

Frank Biermann and Oran R. Young, series editors

Related Books from Institutional Dimensions of Global Environmental Change: A Core Research Project of the International Human Dimensions Programme on Global Environmental Change

Governing Complex Systems

Social Capital for the Anthropocene

Oran R. Young

The MIT Press
Cambridge, Massachusetts
London, England

This book was set in Sabon by Toppan Best-set Premedia Limited. Printed and bound in the United States of America.

Library of Congress Cataloging-in-Publication Data is available.

ISBN: 978-0-262-03593-4 (hardcover); 978-0-262-53384-3 (paperback)

10 9 8 7 6 5 4 3 2 1

Contents

Series Foreword

Humans now influence all biological and physical systems of the planet. Almost no species, land area, or part of the oceans has remained unaffected by the expansion of the human species. Recent scientific findings suggest that the entire earth system now operates outside the normal state exhibited over at least the past 500,000 years. Yet at the same time, it is apparent that the institutions, organizations, and mechanisms by which humans govern their relationship with the natural environment and global biogeochemical systems are utterly insufficient—and poorly understood. More fundamental and applied research is needed.

Such research is no easy undertaking. It must span the entire globe because only integrated global solutions can ensure a sustainable coevolution of biophysical and socioeconomic systems. But it must also draw on local experiences and insights. Research on earth system governance must be about places in all their diversity, yet seek to integrate place-based research within a global understanding of the myriad human interactions with the earth system. Eventually, the task is to develop integrated systems of governance, from the local to the global level, that ensure the sustainable development of the coupled socioecological system that the Earth has become.

The series Earth System Governance is designed to address this research challenge. Books in this series will pursue this challenge from a variety of disciplinary perspectives, at different levels of governance, and with a range of methods. Yet all will further one common aim: analyzing current systems of earth system governance with a view to increased understanding and possible improvements and reform. Books in this series will be of interest to the academic community but will also inform practitioners and at times contribute to policy debates.

This series is related to the long-term international research program "Earth System Governance Project."

Frank Biermann, Copernicus Institute of Sustainable Development, Utrecht University
Oran R. Young, Bren School, University of California, Santa Barbara
Earth System Governance Series Editors

Preface

In this book, I seek to produce insights about governing complex systems by joining two substantial streams of thought, each a continuing focus of interest since the early stages of my career. One stream deals with the analysis of systems and especially the coupled systems that arise in interactions between humans and the biophysical settings in which they operate. The other centers on the study of the social institutions (both formal and informal) that humans devise in their efforts to steer interactions among themselves and between themselves and the biophysical environment. Combining the two streams of thought, I argue, provides a basis for articulating a way of thinking that can help us to understand and respond to the challenges of governing complex systems in the Anthropocene.

As a graduate student at Yale University in the early 1960s, I became intrigued with the power of systems analysis as a way of framing important questions and encouraging the development of rigorous methods applicable to human phenomena as well as biophysical phenomena. There, I worked under the guidance of Robert Dahl, who had recently published his landmark study of political systems entitled *Who Governs?* (Dahl 1961) and Karl Deutsch, who was just finishing his major theoretical work on political systems entitled *The Nerves of Government* (Deutsch 1963). During that time, I also encountered the work of scholars, such as Ludwig von Bertalanffy, who operated under the banner of General Systems Theory and joined forces in an effort to develop a set of concepts (e.g., entropy, feedback, homeostasis, equifinality) of sufficient generality to be applicable to a wide range of systems across the social as well as the natural sciences. The result was an abiding interest in systems thinking as a means of building bridges between the social sciences and the natural sciences.

For various reasons, I set this program aside in the process of writing a dissertation and several early books dealing with bargaining and negotiation as processes through which humans seek to come to terms with social conflicts. But I never lost interest in what my Dartmouth colleague Dana Meadows liked to call "thinking in systems" (Meadows 2008). My engagement in the global environmental change research community, starting in the 1980s, provided an ideal opportunity to rekindle this interest in systems and systems analysis. Now, we find ourselves seeking to understand the dynamics of coupled human-natural systems in which anthropogenic forces are increasingly important and, in many cases, emerging as dominant drivers (Vitousek et al. 1997; Steffen et al. 2004). Here is a class of systems whose complexity exceeds anything familiar to those focusing on biophysical systems or social systems alone. And it has become increasingly clear in the course of time that understanding these coupled systems will be essential to the pursuit of sustainability on scales ranging from the local to the global.

In the 1960s and the 1970s, while working first at Princeton University and then at the University of Texas (Austin) and blessed with a group of gifted graduate students, I began to think about social institutions and the roles they play in steering societies toward socially desirable outcomes and away from outcomes that are socially undesirable. In most settings, we think of government in the conventional sense of the term as the principal mechanism for handling this steering function. But because I was interested in international society, where there is no government in the ordinary sense, it was natural to look for other mechanisms capable of steering large-scale systems. And the study of international organizations, dominated at the time by a kind of formalistic public administration perspective, did not offer an appealing way forward. This gave rise to my interest in what we now think of as international regimes and, in time, in the development of the idea of "governance without government" (Krasner 1983; Rosenau and Czempiel 1992). The essential thought here is that institutionalized cooperation is possible even in the absence of a government. Nor is the creation of a government sufficient to solve many problems featuring a need for governance. Governments may even become a part of the problem rather than a part of the solution in addressing some needs for governance. This line of thinking also gave rise to a long-standing friendship with Elinor Ostrom. Though her empirical work featured small-scale systems in contrast to my work on large-scale systems, we soon realized that we had a great deal in common in thinking about the conditions under which governance without government can succeed, and an unending topic of conversation in thinking about the extent to which propositions about governance scale up and scale down across levels of social organization (Ostrom 1990; Keohane and Ostrom 1995; Young 2005a).

Connecting the focus on systems and the focus on institutions initiated my work on governance and governance systems, now spanning some four decades. And the rise to prominence in the 1970s of the environment as an increasingly prominent policy arena made it natural to bring my theoretical concerns to bear on emerging needs for environmental governance (Young 1982a). As it turns out, this was a fortunate choice. Not only have environmental issues become increasingly prominent on an international or global level; some of the most significant successes in the practice of governance without government have occurred in efforts to solve large-scale environmental issues, such as the depletion of stratospheric ozone and the control of long-range transboundary air pollution. The result has been an unusual cross-fertilization between the communities of policy-makers and scholars seeking to devise social institutions capable of meeting needs for governance at the international and increasingly the global scale.

Now, we are confronted with new challenges arising in conjunction with human-dominated systems on a global scale. Through a combination of population growth, increased affluence, and technological advances, human actions have become increasingly powerful drivers on a planetary scale. We now think increasingly in terms of concepts like "planetary boundaries" and ask ourselves whether we are capable as a species of taking the steps needed to ensure the maintenance of a "safe operating space for humanity" (Rockström et al. 2009; Harari 2015). In effect, we are operating in a world of increasingly complex systems; the need for governance to maintain the sustainability of these systems has grown proportionately. This book is about the challenge of coming to terms with this situation. There are no easy solutions. Many efforts to respond to this challenge prove abortive. But the fate of the human enterprise will be determined in a profound sense by our efforts to become more skillful in governing complex systems.

I owe a profound debt in thinking about these issues to the colleagues I have worked with in the global environmental change research community. This engagement began in the 1980s when I accepted an invitation to serve as the founding chair of the Committee on the Human Dimensions of Global Change in the US National Academy of Sciences (Stern, Young, and Drukman 1991). It continued through my leadership of the international project on the Institutional Dimensions of Global Environmental Change (IDGEC), my co-chairmanship of the Global Carbon Project (GCP), my participation in the Earth System Science Partnership (ESSP), my chairmanship of the scientific committee of the International Human Dimensions Programme on Global Environmental Change (IHDP), and my active engagement in the effort to integrate the major components of the global environmental change research community into what we know now as Future Earth (Young et al. 1999/2005; Young, King, and Schroeder 2008). While I have stepped back from leadership roles in this community, I continue to benefit enormously from my engagement with the Earth System Governance (ESG) project. ESG is, in many ways, a successor to IDGEC, and it has had the good fortune of being guided by an exceptional group of younger scholars (Biermann et al. 2009). The people I have worked with in all these endeavors are too numerous to identify individually. But together they constitute an intellectual community that has stimulated and challenged my thinking over a period now spanning thirty years. My gratitude to the members of this community is deep and lasting.

I do want to express particular thanks to several individuals. Beth Clevenger, my editor at MIT Press, and Frank Biermann, my colleague in guiding the Earth System Governance book series at MIT Press, read an earlier draft of this book and offered candid and constructive advice that allowed me to improve the final product substantially. Three anonymous reviewers provided constructive input, provoking me to think again about several substantive issues and saving me from a number of specific errors. Senior scholars are in the habit of doling out advice; they are less often recipients of candid feedback. But they are as much in need of good advice as others.

Although it has deep intellectual roots, all the work included in this book reflects ideas I have been wrestling with during the 2010s. The individual chapters constitute a blend of new papers prepared specially for the book and previously published papers revised substantially for purposes of analyzing the challenges of governing complex systems. I have endeavored throughout to integrate this work into a whole that is greater than the sum of the parts. My goal is to stake out a strong vision that can serve as a road map guiding the work of analysts and policy-makers seeking to build the social capital needed to promote sustainability in a world of complex systems.

An earlier version of chapter 1 appeared as part of a special issue of *International Environmental Agreements* (2013) devoted to my work and edited by Ronald B. Mitchell. Chapter 2 contains the principal sections of a paper I published in *The Proceedings of the National Academy of Sciences USA* in 2011. Chapter 4 is a revised version of a chapter included in a volume entitled *Global Environmental Commons,* edited by a team of European scholars led by Eric Brousseau (2012). Chapter 5 is a revised version of a paper prepared for a volume of essays edited by Norichika Kanie and Frank Biermann on the UN's Sustainable Development Goals (2017). An earlier version of chapter 6 appeared in a volume on *Ethics and International Affairs* edited by Jean-Marc Coicaud and Daniel Warner (2013).

Introduction: The Age of Complexity

One of the most powerful and robust findings to emerge from the contemporary literature on governance centers on what we have come to think of as the "problem of fit" (Young et al. 1999/2005; Young 2002; Galaz et al. 2008; Young, King, and Schroeder 2008). The essential insight is that the effectiveness of governance systems addressing issues of conservation, environmental protection, and, more generally, sustainability depends on the extent to which the attributes of the relevant institutions are well matched to the properties of the biophysical and socioeconomic systems they are intended to steer or guide. A good fit between these attributes and properties is not sufficient to ensure that governance systems will prove effective or successful. It is easy enough to find examples of situations in which the fit seems good at least on paper but the relevant institutions fail to make the transition from paper to practice in a manner that ensures success. But there is compelling evidence to support the proposition that a good match is a necessary condition for success. As in many other realms, one size does not fit all. That is why many of us who work in this area have concluded that we need to develop a practice of institutional diagnostics that allows us to go beyond panaceas or dogmatic prescriptions in an effort to create and implement governance systems designed specifically to meet the challenges arising in concrete situations (Young 2002; Ostrom et al. 2007; Young 2008).

In some cases, it is easy to identify what is needed to achieve a proper match between the attributes of a governance system and the properties of the natural, social, or socioecological system it seeks to steer. A regime designed to protect highly migratory species (e.g., many species of wild animals and birds) that does not cover the full range of their migratory cycles will be poorly equipped to achieve its goals. A governance system intended to ensure sustainable harvesting of a species straddling the jurisdictional boundaries separating two or more independent states that does not include mechanisms to promote coordination across these jurisdictional boundaries will be unable to coordinate the management practices of the states in question in a manner leading to sustainable yields in harvesting the relevant species. An arrangement designed to protect marine ecosystems that fails to address external threats, such as runoffs from chemical fertilizers and pesticides used in (often) distant agricultural operations but causing hypoxia or dead zones in the relevant marine ecosystems, will be unable to ensure the integrity of these systems.

These examples are clear and compelling; they serve to illustrate the basic insight associated with the problem of fit in an unambiguous manner. Yet all these examples deal with relationships that are comparatively simple and, for the most part, well understood, even though those responsible for creating governance systems to address them may find it difficult to act on the relevant insights in specific cases. Increasingly and especially with the onset of what we have come to think of as the Anthropocene, however, we are encountering the challenge of comprehending and addressing the problem of fit in a complex setting featuring human-dominated systems that behave in ways that are difficult to grasp clearly, even with advanced methods of data collection and analysis and sophisticated modeling procedures (Steffen et al. 2011; *The Economist* 2011). As Raskin and colleagues noted in a prescient observation from 2002, the "human project" has reached a planetary scale, a development that led them to focus on what they called "... our stunningly complex planetary system" (Raskin et al. 2002, 6, 13). The course of human-environment relations in the intervening years has served to reinforce this observation.

The defining feature of the Anthropocene is that human actions have become major drivers of the dynamics of the Earth system (Galaz 2014; McNeill and Engelke 2014; Rockström and Klum 2015; Steffen et al. 2015; Davies 2016; Matson, Clark, and Andersson 2016; Waters et al. 2016). Among the most important of these anthropogenic forces are the variables included in the familiar I = PAT formula, where I stands for impact, P for population, A for affluence, and T for technology (Ehrlich and Holdren 1971). In a world of over 7 billion people, climbing to 9+ billion within the foreseeable future, the biomass of humans is a major factor on a planetary scale, regardless of their level of consumption as individuals. When affluence allows increasing numbers of individuals to adopt upscale lifestyles featuring high levels of material consumption, the result is a major source of planetary problems, such as climate change and the loss of biological diversity. For its part, technology cuts in both directions (Commoner 1971; Simon 1981; Sabin 2014). Technology may help to solve some problems. New technologies may allow us to produce energy in ways that minimize the burning of fossil fuels or even bypass the use of fossil fuels altogether. In the event that the impacts of climate change become increasingly severe, geoengineering may make it possible to avoid some of the most disruptive consequences of climate change (National Research Council 2015a and 2015b). On the other hand, technology in such forms as the steam engine, manned flight, and weapons of mass destruction is a force that has accelerated and intensified human dominance of the Earth system. Relying on technology to solve socio-ecological problems (e.g., the use of nuclear power to reduce greenhouse gas emissions, or the deployment of geoengineering measures to offset the effects of increased concentrations of greenhouse gases in the Earth's atmosphere) may generate problems that are at least as severe as those they are intended to solve. Of course, none of this means that biophysical forces are no longer important drivers of the dynamics of the Earth system. But efforts to devise effective governance arrangements in a world of complex systems must reckon with the extent of human domination of the dynamics of the Earth system on a planetary scale.

The thesis of this book is that solving the problems of the Anthropocene will require the creation and operation of innovative steering mechanisms that differ in important respects from those familiar to us from past experience. This does not mean that the mainstream regulatory approach with its emphasis on the formulation of rules, the promulgation of regulations, and the development of compliance mechanisms is no longer useful. But it does mean that we need to expand our toolkit, adding new ways of responding to needs for governance that can serve either as alternatives to the regulatory approach or as supplements to this approach in specific situations. In developing this thesis, the substantive chapters of the book start with a discussion of enduring insights concerning the effectiveness of governance systems, proceed to an examination of the emerging challenges of the Anthropocene, and finish with an analysis of several new approaches to governance that can enhance the social capital available to those seeking to craft effective responses to these challenges. As a final note, I consider the implications of this line of thinking for science and policy and for the interface between the two. To prepare the way for this analysis, I focus in this chapter on the nature of complex systems and the attributes of such systems that are particularly relevant to thinking about the effectiveness of governance systems.

The Nature of Complex Systems

What we have come to characterize as complex systems have a number of features that set them apart from other systems and that pose problems for a business-as-usual approach to thinking about meeting needs for governing human-environment relations. Four principal clusters of features, which often interact with one another, present challenges to those seeking to design and operate governance arrangements capable of producing sustainable results on a large scale: (i) connectivity or tight coupling among system components, (ii) thresholds, triggers, and nonlinear patterns of change, (iii) dynamic and directional processes, and (iv) emergent properties and the frequency of surprises (Levin 1999; Janssen 2002; Johnson 2007; Mitchell 2009; Scheffer 2009).

Much has been written about the rise of connectivity among the biophysical and socioeconomic components of the Earth system. We lack a simple metric to use in tracking the rise of interdependencies over time in a quantitative manner. But three major developments in this realm are critical in thinking about the governance of complex systems. Perhaps the most familiar of the three is captured in the concept of globalization (Held et al. 1999; Young et al. 2006b). In simple terms, globalization means that we are dealing with an Earth system in which it is no longer possible to isolate or insulate developments occurring within spatially delimited areas from the interplay of forces operating in the world at large. A second critical factor has to do with the increasingly tight connections between biophysical systems and socioeconomic systems. As others have observed, we now live in a world of human-dominated systems in which human actions are major drivers of change from the local level to the global level (Vitousek et al. 1997; Steffen et al. 2004; Steffen et al. 2015). This means we need to focus on the behavior of socioecological systems rather than treating socioeconomic systems as subject to the impacts of biophysical forces that humans are powerless to control or manipulate (Gunderson and Holling 2002; Liu et al. 2015). The third development is captured in the concept of teleconnections or telecouplings (Liu et al. 2013). The essential point here is that forces operating in one part of the Earth system can trigger unintended and often surprising consequences that show up in distant parts of the system. A prominent example features the recession and thinning of sea ice in the Arctic arising as a consequence of emissions of greenhouse gases (GHGs) occurring largely in the mid-latitudes, and the resultant positive feedback processes accelerating the loss of sea ice in the Arctic and affecting weather patterns in distant areas. But similar telecouplings are increasingly well-documented in many realms.

Complex systems are characterized by nonlinear behavior of a sort that is difficult to grasp in terms of models of gradual and incremental processes that dominate mainstream thinking about environmental protection and sustainable development. Typical in this connection are the occurrence of thresholds (or tipping points as they are popularly known), trigger mechanisms, and regime shifts or state changes that break the grip of path dependence and catapult a system onto a different trajectory (Gladwell 2002; Lenton et al. 2008; Scheffer 2009). Such processes feature the operation of positive feedback loops and chain reactions leading to the propagation of cascades of change that shift systems from one trajectory to another (often) dramatically different trajectory.

Often, the resultant shifts are irreversible in the sense that there is no prospect of reverting to the status quo ante. Prominent examples involve the propagation of worldwide depressions (e.g., the Great Depression of the 1930s) and large-scale transitions in the planet's climate system (e.g., the current processes leading to climate change and variability at a speed that has no analogue in the last 800,000 years or more). Even when they exhibit regular cycles (often described as oscillations), complex systems do not return to the same starting point following each cycle or oscillation. Compare the annual cycles of the seasons or the school year in this regard with Earth system cycles or major business and political cycles. We expect the cycle of the seasons to occur on a regular basis with little or no change in the equinoxes and solstices; they are products of recurrent and predictable patterns of planetary behavior. We expect one school year to follow another with little change in the nature of the cycle from one year to the next. But this is not the case with global physical processes or large-scale business and political processes. Once the Earth system transitions from the Pleistocene to the Holocene or from the Holocene to the Anthropocene, we expect permanent changes in the nature of the system, at least on any time scale of interest to those concerned with human well-being. Business and political cycles are particularly interesting in this regard. Recoveries from mild recessions may bring the economy back more or less to the status quo ante. Routine electoral cycles may take place in a manner that produces little or no change in the character of the political system. But this is not the case with regard to more dramatic cycles of this sort. The election of 1860 in the United States led directly to the Civil War, with profound and irreversible consequences for the nature of the American political system. The Great Depression starting in 1929 precipitated worldwide economic disruptions, leading to changes in the character of economic and political systems that have proven irreversible ever since.

Increasingly, this means that we are dealing with directional processes or what systems analysts often call bifurcations in contrast to oscillations (Lenton et al. 2008). As Scheffer notes, "social systems never really shift back to a state that they have been in before, as ecosystems … sometimes do" (Scheffer 2009, 240) This feature is even more prominent in human-dominated socioecological systems, a fact that raises basic questions about the comforting implications of the familiar theory of the adaptive cycle with its emphasis on resilience in dynamic settings.[1] This third feature of complex systems ensures that governance strategies that worked in the past may well prove ineffective in the future when essential features of the relevant systems experience changes that have important consequences for the results flowing from policy interventions. As many observers have noted, this makes it hazardous to rely on lessons from past experience in developing plans to meet future needs for governance.

Under the circumstances, it will come as no surprise that we think in terms of the concept of emergent properties in efforts to grasp the behavior of complex systems. Some have argued that the concept of emergence is vague or imprecise, simply reflecting a lack of understanding of the behavior of systems of interest (Mitchell 2009). There is some truth to this observation. But the essential insight here is that some of the systems we care about most have so many interactive elements or components and are so sensitive to initial conditions that we are unable to anticipate the consequences of our efforts to intervene in these systems in the pursuit of human goals or objectives and are often surprised by the actual results (Perrow 1984). As keen observers have noted, predictions regarding the consequences of macroeconomic policy interventions are seldom borne out under real-world conditions (Silver 2012). That is why there is no end to the debates between advocates of monetary policies and fiscal policies and between advocates of fiscal austerity and stimulus spending, whose forecasts regarding the effects of their preferred economic policies are seldom borne out in practice and almost never subject to decisive or definitive tests. An equally dramatic case involves the Earth's climate system and efforts to craft policy interventions that will, in the words of the UN Framework Convention on Climate Change, "prevent dangerous anthropogenic interference with the climate system" (UN Framework Convention of Climate Change, 1992, Art. 2).

This means that surprise is a key feature of the behavior of complex systems. Despite the development of increasingly sophisticated models of the Earth's climate system, known as general circulation models or GCMs, for example, we are frequently taken by surprise by the actual behavior of this system, so much so that it remains difficult to educate even the most informed members of the general public regarding the nature of climate change, and so-called climate deniers are still able to gain a certain amount of traction in claiming that the whole issue of climate change is a hoax. Of course, sophisticated systems analysis may reduce the frequency of surprises in due course, and simulations may help to improve our understanding of the behavior of complex systems. But for the foreseeable future, we must reckon with the fact that steering these systems will require decision-making under uncertainty, a condition that even the most sophisticated forms of risk assessment cannot alleviate. One obvious implication of this situation centers on the need to create governance systems that are nimble or agile in the sense that they can adapt easily to changing circumstances, without losing their capacity to steer human-environment interactions effectively.

Self-Organizing Systems

Systems analysts have observed that some systems exhibit self-organizing properties (Gunderson and Holling 2002; Meadows 2008). From the perspective of governance, the significance of this observation is that such systems behave in a manner that is orderly or sustainable in the absence of any identifiable steering or guidance mechanisms. In such situations, governance may be superfluous, or useful mainly to pursue goals other than sustainability (e.g., preferred outcomes regarding issues of social justice).

It is worth distinguishing among at least three distinct types or forms of self-organization in this connection. One centers on the ability of large numbers of organisms/agents to engage in joint maneuvers or to collaborate in pursuit of common goals in the absence of any conscious decision-making processes. A second type of self-organization involves the achievement and maintenance of a condition of equilibrium in systems involving large numbers of actors who make decisions based on their individual interests and do not engage in conscious efforts to coordinate their individual activities. A third features the idea of resilience in the sense of the capacity of a system to adapt to changing circumstances and undergo internal reorganization without experiencing a change of state or losing its capacity to function effectively. Systems of this sort are marked by an ability to monitor various environmental conditions and to initiate changes needed to maintain a steady state or to ensure that the level of a critical variable (e.g., temperature) remains within an acceptable range through the use of feedback mechanisms and an ability to make timely adjustments in the face of both external and internal pressures.

In some cases, self-organizing behavior appears to be a matter of genetic programming rather than individual choice. Consider the examples of flocks of birds that are able to fly in formation without bumping into one another or even deviating from tight formations in complex maneuvers, or colonies of ants in which the minute contributions of large numbers of individuals can produce elaborate colonial structures (Mitchell 2009). Behavior of this sort is rooted in the genetic programming of the individual members of flocks and colonies and is a product of evolution. There is no indication that it involves anything resembling a cognitive process. The results are impressive so long as the relevant external conditions remain unchanged. But the programmed nature of these forms of self-organization no doubt explains as well why collective action in such systems can collapse when conditions change. Unless the individual organisms in question show some ability to "learn" in the sense of making timely and suitable adjustments in their behavior in response to changes in their environment, the self-organizing character of the relevant systems will be sharply constrained. From a species perspective, evolution may provide an effective learning mechanism over the long haul. But learning of this sort is unlikely to be an effective strategy for operating sustainably in complex systems in which thresholds and trigger mechanisms regularly produce changes that are nonlinear and (often) abrupt.

In other cases, self-organization results from the operation of what systems analysts think of as negative feedback mechanisms (Wiener 1948). Perhaps the classic illustration is the perfectly competitive market that exists in elementary economics textbooks, though it may be exceedingly rare or non-existent under real-world conditions. In such a system, self-interested buyers and sellers engage in interactive decision-making that determines the price of a particular good or service. If rising demand leads to scarcity that drives up prices, additional producers will be motivated to enter the market, and the prices will fall. A dearth of buyers, on the other hand, will lower prices and drive some producers out of the market. Assuming the market is perfectly competitive, equilibrium will occur at a price that clears the market in the sense that every unit of the good or service available finds a willing buyer. This is the basic process captured in the familiar concept of the invisible hand. Systems analysts have applied similar types of models to other phenomena such as competitive arms races (Rapoport 1960).

Of course, even elementary textbooks point to a host of conditions that can give rise to market failures, including the occurrence of natural monopolies, the development of oligopoly, the effects of imperfect information, the difficulties associated with the supply of public goods, the prospect that technological change will make a market obsolete, and so forth. From the perspective of governance, however, it is worth emphasizing that even a perfectly competitive market will not function effectively in the absence of a more or less elaborate structure of rules or institutional arrangements dealing with matters like property rights, contracts, liability, truth in advertising, and compliance.[2] Even under ideal circumstances, therefore, self-organizing markets are not likely to arise, much less to produce socially desirable outcomes, in the absence of a supportive system of institutional arrangements.

It has become fashionable in recent years to focus on the idea of resilience as the key to successful self-organization in complex adaptive systems, especially in settings that feature coupled human/natural systems or what we often characterize as socioecological systems (Gunderson and Holling 2002; Berkes, Codding, and Folke 2003; Folke 2006; Walker and Salt 2006). In the well-known formulation of Holling and the school of thought rooted in his work, some systems exhibit a capacity to reorganize themselves in such a way as to adapt to changing circumstances without losing the ability to perform their normal functions (Holling and Gunderson 2002). A corporation that reorganizes itself to enhance productivity or to take advantage of technological advances that allow for the development of new products provides a simple example. Apple and Google, for example, have developed a capacity to reorganize themselves almost continuously, without losing their identities as firms specializing in information technologies.

But resilience becomes more problematic in complex and, especially, human-dominated systems that feature bifurcations in contrast to oscillations. Consider systems that reach tipping points like the European political system in the summer of 1914 on the eve of World War I, or the global economy in the early months of 1929 on the eve of the Great Depression that began later that year. Both cases presented critical needs for learning and adjustment to avoid systemic collapses or disruptions with far-reaching consequences including an absence of winners. But the necessary learning and adaptation were not forthcoming in either case. Ironically, cases of this sort often involve human systems or human-dominated systems in which it may seem reasonable to expect that the learning needed to maintain resilience will occur (Dryzek 2014). But as these examples make clear, there is no basis for assuming that learning will occur under such conditions in a timely manner needed to prevent catastrophe. Under the circumstances, the appealing idea of the adaptive cycle may well lead us astray as we search for ways to understand the dynamics of human-dominated systems (Gunderson and Holling 2002).

Governing Complex Systems

We can hope that processes of self-organization will prevail in a wide range of biophysical, socioeconomic, and socioecological systems that are important from the perspective of human well-being. Such hopes may be rewarded under some conditions. The idea of an "invisible hand" is helpful in some contexts; what some thinkers, especially those associated with conservative political thought, call "spontaneous order" does occur under some conditions (Hayek 1973). But it does not take a lot of analysis to convince ourselves that relying exclusively on processes of self-organization will be a forlorn hope with regard to many systems we care about, and that this is particularly true in the case of complex systems. Individual markets are subject to dramatic fluctuations, whole economies collapse, political systems disintegrate, and international relationships fluctuate between war and peace. Biophysical systems are subject to nonlinear state changes in which large rivers change course, ice sheets disintegrate, and species go extinct. Nor can we count on the advent of an era characterized by human domination of large socioecological systems to enhance negative feedback mechanisms of the sort that make some large and dynamic systems resilient. As the literature on "planetary boundaries" makes clear, human actions are pushing many large Earth system elements toward conditions in which nonlinear and dramatic change is more and more likely (Rockström et al. 2009; Rockström and Klum 2015; Steffen et al. 2015). The climate system is a dramatic example of a complex system that is changing dramatically under the weight of anthropogenic forces (Archer and Rahmstorf 2010). But it is far from the only case in which such processes are evident.

This is where the need for governance comes into focus as a prominent concern. If we treat governance as a social function centered on steering societies toward collectively desirable outcomes and away from collectively undesirable outcomes, it becomes apparent that success in creating and implementing governance systems is a critical determinant of sustainability in the pursuit of human well-being in many settings (Young 1999). This does not mean that we must turn to governments in the conventional sense to meet needs for governance arising in a variety of settings. We know now that "governance without government" is an important phenomenon in many social settings (Rosenau and Czempiel 1992). First identified and analyzed clearly in small-scale settings in which humans develop informal rules and social practices to promote sustainability in the use of natural resources (Ostrom 1990), we now understand that the institutions needed to achieve governance without government can occur at other levels of social organization as well (Dietz, Ostrom, and Stern 2003). At the international level, for example, where there is little likelihood of the emergence of a government in the conventional sense, humans have developed more or less effective institutions to address needs for governance relating to the dismantling of barriers to free trade, the development of prohibitions on the use of chemical weapons in warfare, and the imposition of restrictions on the production and consumption of ozone-depleting substances. Clearly, there is no guarantee of success in such efforts. Just as economists spend a great deal of time seeking to understand and deal with market failures, those who think about governance more generally focus a lot of attention on factors affecting the creation of governance systems, the determinants of the success or effectiveness of these systems, and the prospects for repairing or replacing these systems when they fail to produce socially desirable outcomes. Governance failure, treated as the counterpart of market failure, is an important phenomenon worthy of sustained analytic attention (Wolf 1988). An effort to identify the determinants of success and failure in the operation of governance systems is a major focus of attention in the contemporary literature on governance (Young 2011a).

These observations are generic; they apply to efforts to meet needs for governance under all circumstances. What are the consequences of the advent of complex systems on a global scale in this context (Duit and Galaz 2008)? Over time, I have come to the conclusion that the generic types of problems giving rise to needs for governance are the same in complex systems as they are in most other systems. But the challenge of meeting these needs rises rapidly as systems become more complex. To understand the implications of this observation, consider needs for governance involving: (i) overcoming collective-action problems, (ii) internalizing social costs or externalities, (iii) limiting path dependence, and (iv) coping with risk and uncertainty.

Collective-action problems are familiar at all levels of social organization. To use Schelling's terminology, they arise when micromotives produce macrobehavior that is collectively undesirable in the sense that no one benefits from the outcomes (Schelling 1978). Situations involving the tragedy of the commons and widespread incentives to behave as free riders with regard to the provision of public goods provide well-known examples (Olson 1965; Hardin 1968). The source of the problem lies in the fact that individuals have strong incentives to ignore the need to act cooperatively, whether this means refusing to limit their use of a common pool resource or to contribute to the cost of supplying a non-excludable public good. While the classic examples feature small-scale systems, large-scale problems like the provision of clean air on a regional or even global scale give rise to the same pathology.

What makes problems of this sort unusually difficult to solve in complex systems has to do with their scale, nonlinear features, and tendency to generate abrupt and surprising developments. The problem of climate change exemplifies these properties. We cannot solve the problem of climate change piecemeal. The efforts of one country or even a multinational region acting alone will not make much difference. What is needed is a concerted and persistent effort on the part of all the major players in the system to reduce drastically their emissions of greenhouse gases. But while we sometimes speak of the "international community" in discussions of such matters, there is no history of social solidarity to draw on at this level (Hurrell 2007). We cannot count on the influence of feelings of guilt or shame, shared experiences, or an established culture of cooperation to facilitate efforts to solve collective-action problems on a large scale. What is more, the Earth's climate system encompasses a number of tipping elements. It is entirely possible that we will reach a threshold at which some seemingly modest trigger ignites a cascade of nonlinear events that produce dramatic and surprising shifts in the climate system in a relatively short period of time (Cornell et al. 2012). What this means is that we may not have the leeway to experiment with different solutions, searching on a trial-and-error basis for an effective means to address the problem over a lengthy period of time. At the same time, the climate system is immensely complex. The behavior of this system regularly takes us by surprise. Sometimes the surprises seem benign, as in the case of the recent (apparent) pause in the rate of increase in temperatures at the Earth's surface.[3] But there is no guarantee that this will be the case in the future. Climate surprises in the future may take the form of exogenous shocks that are difficult for even the most affluent societies to cope with effectively and that literally cause the disintegration of less affluent societies. Knowing that surprises are likely to occur but being unable to predict what form they will take or when they will occur adds enormously to the challenge of solving collective-action problems in complex systems.

Likewise, there is nothing new or unfamiliar about problems featuring social costs or what economists refer to as externalities. Social costs are unintended (and often unforeseen) side effects of actions taken in pursuit of more or less well-defined (and normally legitimate) goals or objectives. In the typical case, the prevailing rules of the game do not provide the relevant actors with strong incentives to avoid or curb their actions. Traditionally, social costs have been treated as relatively modest or marginal concerns. Legal concepts like the nuisance doctrine have evolved to address these problems in small-scale settings. Cases in which cows get loose and damage the neighbor's garden or sparks from a coal-fired locomotive damage nearby property are well-known examples. For their part, economists have argued that the parties themselves may well be able to solve such problems through direct negotiations (Coase 1960). They are fond of noting as well that the consequences of some externalities are positive rather than negative, as in cases where innovations catalyze the development of new products that improve the quality of life for large groups of consumers or even whole societies.

In large and complex systems, however, the scope of such problems alters their significance profoundly. Climate change is fundamentally a problem of social cost. No one intends for their emissions of greenhouse gases to produce disruptive impacts on the Earth's climate system. The rules of the game simply fail to force emitters to internalize the costs associated with the impacts of their actions on the climate system. In effect, emissions of such gases have long been treated as cost-free waste streams, thereby reducing or even eliminating incentives to curb them. In complex systems, moreover, it is difficult to prove beyond a reasonable doubt that the activities of identifiable actors cause the impacts in question. Court cases addressing what are often called toxic torts are replete with examples of the challenges involved in efforts to link the actions of specific polluters definitively to outcomes in such areas as human health (Harr 1996; Fagin 2013). The prevalence of telecouplings and nonlinear processes adds to the problems of coping with the consequences of social costs in large and complex systems. The impacts of externalities may show up in distant places and in a manner that takes even the keenest observers by surprise. The prospect that a regime shift will occur once a threshold or tipping point is passed simply adds to the problems of assigning responsibility to the actions of individual polluters. What was once a limited concern subject to management through the use of ideas like the nuisance doctrine has now become a global phenomenon with respect to which effective regulation would require far-reaching changes in the rules of the game operating at the level of international society.

A related problem centers on what we commonly think of in terms of the idea of path dependence or the tendency of systems to proceed along predetermined paths regardless of the impacts of external forces. Path dependence is not in itself a bad thing. But there are many cases in which path dependence leads to collectively undesirable outcomes, even when those engaged in the relevant actions are conscious of the dilemma. Consider examples like the problem of stopping the mad momentum of arms races, where individual participants are unable to break the action/reaction dynamic leading to higher and higher levels of armaments; the challenge of shifting course once players in a system have invested in facilities (e.g., coal-fired power stations) that need to be amortized over a long period of time; the problem of overcoming the lock-in effect associated with the creation of quasi-property rights (e.g., individually transferable quotas or ITQs) in marine fisheries; or the difficulty of interrupting the competitive dynamic between computer programmers seeking to build secure systems and hackers who spend their time devising ways to crack the latest inventions of the programmers. The basic point is simple. It often requires more than an awareness of a problem to overcome the influence of the dysfunctional behavior that causes it. One of the classic functions of governance systems is to find ways to avoid or overcome the straitjacket of path dependence, despite the opposition of blocking coalitions that can complicate or even stymie such efforts in concrete cases.

Here, again, the problem becomes more severe in the context of complex and dynamic systems, even though the underlying mechanism at work is readily recognizable. As before, the case of climate change provides a clear illustration of this proposition. As many observers have noted, our fossil fuel–based economies are highly path dependent. Producers of coal, oil, and gas have not only made huge investments in extraction facilities and transportation systems; they also benefit from large public subsidies and wield enormous influence in many political arenas. Similarly, many electricity producers are stuck with long-term investments in coal-fired power stations or long-term contracts with suppliers of coal. Some are beginning to switch to gas-fired plants, but this does not offer a long-term solution to the problem. Most of those who have analyzed the situation carefully have concluded that we need transformative change in which modern economies both reduce their energy demands and rely increasingly on renewable sources of energy (e.g., solar, wind, hydro power). There are good reasons to believe that it is possible to make such a transition without disrupting the economic foundations of modern and postmodern societies. Once we turn the corner in this effort, in fact, many entrepreneurs almost certainly will discover new opportunities to profit from this transformative process. But given the force of path dependence, initiating and guiding such a transformation is unlikely to occur in the absence of far-reaching adjustments in the rules of the game favoring the status quo.

Analogous observations are in order regarding the problem of uncertainty.[4] There is nothing unusual about the need to make decisions under uncertainty and, as a result, to commit financial and other material resources to initiatives whose chances of succeeding are difficult or impossible to calculate. So long as risks are not extreme and the resources involved are limited, it is reasonable to expect private actors to take the lead under conditions of this sort. Such situations will produce both winners and losers, but this need not lead to systemic problems. When both the riskiness and the scale of the resources involved reach a certain level, however, public support becomes essential. This is particularly true in the case of goods that have some of the features of public goods and especially when these goods are lumpy, so that the initial cost of supplying such goods is high even though the marginal cost of supplying them to additional users is low. This is why large-scale infrastructure projects often require some form of public funding. As the case of toll roads suggests, shrewd entrepreneurs may find ways to create exclusion mechanisms that allow them to make a profit from supplying goods even in such cases. Still, some form of public support often is needed to launch initiatives of this kind. Think of the origins of the Internet, a service we now take for granted, which owes a lot to American defense spending, as a case in point.

The transition to large and complex systems simply intensifies these concerns. The combination of teleconnectons or telecouplings, nonlinear processes, and emergent properties increases uncertainty to levels with which we have little experience. Despite the advent of big data, the development of supercomputers, and increasing reliance on more sophisticated models, we are routinely asked to make decisions in situations where we have little capacity to anticipate the results. When planetary systems are involved (e.g., the carbon cycle, the nitrogen cycle), the scale of the relevant activities is apt to exceed the capacity of the largest private actors, even in cases where benefits eventually exceed costs by a substantial measure. To take the case of greenhouse gas emissions again, we can hope that ordinary entrepreneurs will find ways to profit from the production of solar and wind power, the introduction of cars powered by electricity or hydrogen, and the installation of technologies that allow for carbon capture and storage. But the fact is that developments of this sort on the scale needed to solve the problem of climate change are unlikely to emerge in the absence of fundamental changes in the rules of the game needed to produce far-reaching shifts in incentives. This does not mean that the creation of large state-owned enterprises will be needed to take on such high-risk ventures. But it does mean major changes in institutional arrangements that control incentives are likely to be a necessary part of changes of this sort.

Running through all these observations is the importance of striking a balance between predictability and durability on the one hand and nimbleness or agility in adjusting to change on the other. An institution that has no staying power or capacity to resist pressures to change will have little or no impact on the behavior of those subject to its provisions. Subjects who know that governance systems are extremely responsive to pressures for change will have little reason to adjust their behavior to comply with the provisions of such systems. At the same time, governance systems must be able to adapt to changes in the needs for governance and in the underlying biophysical and socioeconomic conditions that give rise to these needs. This is especially important in the case of dynamic systems where changes are often nonlinear, rapid, and irreversible. Finding effective ways to pursue durability and nimbleness or agility at the same time is critical to the achievement of sustainability in the Anthropocene.

The Architecture of the Book

Making use of an analytical technique often characterized as institutional diagnostics (Young 2002; Ostrom et al. 2007; Young 2008), I show in the chapters to follow that the familiar regulatory approach to governance has serious drawbacks as a means of dealing with a range of increasingly prominent problems arising in complex systems. This leads me to initiate the process of identifying alternative approaches that may prove more effective as tools for achieving sustainability in the Anthropocene. My goal is to contribute to the growth of social capital by thinking systematically about alternatives, rather than to reach definitive conclusions about the best way to proceed. Other approaches are certain to come into focus as we gain experience in this realm. Moreover, while institutional innovation is essential, a good deal of existing knowledge about the governance of large systems remains relevant. A key challenge going forward, then, will be to join together enduring insights regarding governance and new ways of thinking to develop the social capital needed to come to terms with the needs for governance arising in complex systems (Young et al. 2006b; Brondisio, Ostrom, and Young 2009). My contribution is to launch us on this journey rather than to ensure that we make it safely to our destination. Others can and will make major contributions as we explore various avenues that come into focus during the course of this journey.

The architecture of the book is straightforward. In the substantive chapters, I organize my thinking about this subject into three sections labeled "Enduring Insights from Ongoing Research," "Challenges of the Anthropocene," and "New Perspectives on Governance." In the conclusion, I discuss research opportunities, lessons for policy-makers, and life on the science/policy interface in the light of my analysis of governing complex systems. My goal throughout is to land squarely in what analysts have called Pasteur's Quadrant, exploring ideas that, at one and the same time, constitute insights of considerable importance in theoretical terms and have far-reaching practical implications (Stokes 1997; Matson, Clark, and Andersson 2016).

Part I, devoted to the theme of "enduring insights from ongoing research," is based on the premise that there is no need to start from scratch. A good deal of what we have learned over the last several decades in analyzing the creation, effectiveness, and dynamics of regimes dealing with human-environment relations remains relevant as a point of departure for this inquiry regarding the governance of complex systems. The chapters in this part seek to tease out these enduring insights. Chapter 1 starts by introducing a set of concepts and definitions I have developed over the last four decades in work focused on environmental governance but equally relevant to this analysis of governing complex systems. It provides an overview of what we have learned in this field of study that is intended to serve as a launching pad for new and often critical thinking rather than as a statement of irrefutable propositions. Chapter 2 then drills down on the concept of effectiveness, reflecting on what we mean by effectiveness with respect to governance systems and what we have learned about the determinants of success and failure in the operation of international regimes. Measuring effectiveness, much less pinpointing the determinants of success and failure, is analytically challenging; it leads more readily to explanations for the failure of governance systems under various conditions than to simple recipes for success. It also brings us face to face with the phenomenon of complex causality. Nonetheless, there is no getting around the need to address this challenging issue systematically, if we are to make progress in governing complex systems.

Part II, on "Challenges of the Anthropocene," then turns the spotlight directly to the emerging challenges of governing complex systems. The two chapters in this part seek to pinpoint new needs for governance that differ from or go beyond the needs familiar from past experience. The point of departure for chapter 3 is the proposition that thresholds and trigger mechanisms leading to nonlinear changes occur in socioeconomic systems just as they do in biophysical systems, despite the fact that human systems are reflexive in a manner that has no direct analogue in biophysical systems. When we turn to socioecological systems of the sort that are at the heart of discussions about sustainability in the Anthropocene, the interactions between thresholds and trigger mechanisms in biophysical and socioeconomic systems become a critical focus of attention. Among other things, this makes an understanding of bifurcations in contrast to oscillations an increasingly central concern (Scheffer 2009; Brousseau et al. 2012). Chapter 4, on the sustainability transition, explores how these conditions alter the needs for governance familiar to us from work of the sort discussed in part I. This chapter also begins the process of thinking about ways to respond to these challenges, identifying a set of prudent practices available to those responsible for meeting such challenges.

Part III, entitled "New Perspectives on Governance," includes three chapters that identify and explore innovative ways to think about governance that show promise of becoming growth areas in meeting needs for governance arising in the Anthropocene. The goal is to open up this topic, taking some initial steps toward adding to our social capital in this realm rather than to exhaust the range of possibilities. Other new approaches will come into focus as we bring our intellectual resources to bear on the pursuit of knowledge about the governance of complex systems. Perspectives that envision increasingly important roles for a variety of nonstate actors may be particularly important in this regard. Taken together, the results will enhance the social capital available to those seeking to meet needs for governance in the Anthropocene (Brondisio, Ostrom, and Young 2009).

Chapter 5 explores the insight that goal-setting in contrast to rule-making can serve as a governance strategy, offering an approach that provides a way forward in responding to some emerging needs for governance (Kanie and Biermann 2017). The strategy of goal-setting sometimes comes into focus as a complement to rule-making rather than as an alternative. But the idea of goal-setting is interesting in its own right because it directs attention to perspectives on sources of behavior that differ from those embedded in mainstream thinking about rule-making. Chapter 6 turns to what I call principled governance, examining the role of sets of normative principles or ethical systems in the operation of governance systems. Although early studies of international regimes treated principles and norms as elements of regimes, much of the recent literature on governance focuses more on decision-making procedures, rules, and compliance mechanisms (Krasner 1983). Yet it may well be that the influence of what we now think of as the logic of consequences has led us to overemphasize the role of rules, disregarding the power of principles in addressing matters of governance in complex systems (March and Olsen 1998). Chapter 7, then, focuses on the question of the contributions of what is commonly referred to as good governance. Some commentators simply subsume effectiveness under the broad and rather imprecise umbrella of the idea of good governance, a practice that eliminates the issue of considering the links between good governance and the effectiveness of governance systems by definition. Among those who separate the two, however, there is a pronounced tendency to assume that what we generally regard as good governance is a critical determinant of effectiveness. This chapter takes a hard look at this assumption and asks whether good governance can play a role in governing complex systems, whatever its attractions in normative terms.

The conclusion on "Science, Policy, and the Governance of Complex Systems" extracts a set of practical implications from the preceding analysis, not only for scientists and policy-makers operating in their own domains, but also for the interface between science and policy in efforts to meet needs for governance arising in complex systems. How can we develop usable knowledge that is relevant to governing large complex systems? What implications should policy-makers focus on as they seek to come to terms with the dynamics of complex systems? Are there tools that can assist policy-makers as they evaluate options in such settings? Can analysts and policy-makers work together constructively to pursue sustainability in the Anthropocene?

Drawing on my own experience in this realm, I make the case that it is inappropriate to start with a sharp dichotomy between analysis and praxis. Over time, I have come to regard this practice as a hindrance to constructive thinking about the nature of the worlds of analysis and of policy-making and implementation and the interactions between the two. In my own work, I have found that interactions between these worlds can produce benefits running in both directions. My theoretical work on governance has produced insights that have allowed me to pursue constructive activities in several policy arenas (Young 2016). Conversely, encounters with real-world issues have sharpened my thinking about theoretical questions pertaining to governance immeasurably. I have come away from this experience with a firm conviction that operating in Pasteur's Quadrant is not only feasible; it also can enrich both sides of the science/policy interface. To me, this is an appealing way to think about the now fashionable idea that we should embrace the co-production of knowledge (Jasanoff 2004).

I
Enduring Insights from Ongoing Research

1

Environmental Governance as a Point of Departure

Introduction

If we distinguish governance as a social function from government as an organization or assemblage of organizations created under various circumstances to perform the function of governance, it makes sense to ask whether there are conditions under which governance without government is a realistic option. In recent decades, a sizable body of literature devoted to this topic has emerged. A large portion of this literature deals with small-scale systems in which culturally embedded practices serve to coordinate and steer the actions of users or appropriators of natural resources effectively in communities that have no government in the ordinary sense of the term (Ostrom 1990; Ostrom et al. 2002). The research of those joined together under the umbrella of the International Association for the Study of the Commons is particularly noteworthy in this connection.[1] But there are also cases in which governance without government is successful on a much larger scale, despite the fact that it is difficult to identify authentic analogs to culturally embedded social practices operating at the international or global level.

To be sure, governance failures in large-scale settings are common, as they are in all efforts to address needs for governance. There is no basis for taking it for granted that humans will find solutions to the challenges of governance arising in the Anthropocene. A critical concern, then, is identifying the determinants of success in efforts to solve problems like achieving sustainability under the conditions we face today. Because much of the relevant literature deals with environmental issues, I turn to experience in this domain as a point of entry into the overarching theme of governing complex systems, seeking to provide a foundation on which to ground my inquiry into emerging challenges in the chapters of part I and the analysis of new approaches to governance that I develop in the chapters of part III.

The discussion in this chapter proceeds in three steps. The initial substantive section provides a collection of working definitions that have informed my thinking about governing human–environment relations over a number of decades (Young 2012b; Young 2013a). The next section formulates and elaborates on a set of general observations about environmental governance applicable in all settings. The concluding section introduces five insights that focus more specifically on environmental governance at the international or global level. The analysis in chapter 2 builds on this account to provide a more focused assessment of the factors that determine the effectiveness of governance systems.

Working Definitions

I am a nominalist when it comes to definitions. I do not believe there are correct (or essential) definitions of key concepts waiting to be discovered, refined, and adopted by all members of the relevant research community. We are all free to attach whatever meanings we choose to key concepts. Nonetheless, it is essential to use concepts precisely and consistently.[2] A remarkable proportion of apparent disagreements among students of governance and, for that matter, those endeavoring to illuminate any important phenomenon, turn out to be simple matters of definitional imprecision or inconsistency. Once we clarify our definitions, many of these apparent differences evaporate. So, here is a set of definitions covering the most important terms I use in formulating the specific propositions discussed in the next two sections of this chapter and, more generally, in thinking about the challenge of governing complex systems.

Governance is a social function centered on steering collective behavior toward desired outcomes and away from undesirable outcomes. A critical feature of this way of thinking about governance is that it allows us to decouple the idea of governance from the more familiar concept of government. As a result, we can think about a variety of mechanisms through which societies fulfill the function of governance and consider the prospect of providing governance without government, an idea of special importance for those of us focusing on international/global society where there is no government in the ordinary sense of the term. *A governance system is an ensemble of elements performing the function of governance in a given setting. Institutional arrangements form the core of such a system, but the ensemble normally includes cognitive, cultural, and technological elements as well.* Governance systems come in many forms; a key concern in specific settings involves matching governance systems with the issues they address rather than seeking to evaluate the relative merits of alternative governance systems in generic terms.[3]

Institutions are collections of rights, rules, principles, and decision-making procedures that give rise to social practices, assign roles to the participants in these practices, and guide interactions among the participants. Institutions are prominent features of governance systems at all levels of social organization. *Regimes are institutions specialized to addressing functionally defined topics (e.g., health care, pollution, trade) or spatially defined areas (e.g., Antarctica, the North Pacific, Western Europe)* (Young 1982). Regimes constitute a proper subset of institutions. All regimes are institutions, but not all institutions are regimes.

Environmental and resource regimes are institutions that address matters of governance pertaining to human-environment relations. These regimes address questions of governance involving human uses of natural resources (e.g., stands of trees, stocks of fish, deposits of minerals) and human actions affecting biophysical systems (e.g., air and water pollution).[4] Desired outcomes in such settings center on the achievement of sustainability and resilience but typically include matters of efficiency and equity as well. Individual environmental and resource regimes may exhibit both functional and spatial characteristics (e.g., the clean air regime applicable to the United States). When these regimes deal with matters that are transboundary in nature (e.g., the regime dealing with air pollution in Europe) or that lie beyond the jurisdiction of nation states (e.g., the regime for the deep sea bed, the regime for the Earth's climate system), it is standard practice to refer to them as *international environmental regimes*.

An important distinction regarding environmental and resource regimes separates those that are *constitutive* in nature (e.g., the overarching framework for ocean governance set forth in the 1982 UN Convention on the Law of the Sea) from those that are *operational* in nature (e.g., the regulatory measures and decision-making procedures articulated in the 1987 Montreal Protocol on the control of ozone-depleting substances). Constitutive arrangements often provide a foundation for the development of a number of operational regimes; operational regimes typically are nested into the broader frameworks provided by constitutive arrangements. The law of the sea, for instance, underlies a variety of operational arrangements, articulated in implementing agreements or in regimes established by intergovernmental organizations (e.g., the International Maritime Organization) dealing with commercial shipping, industrial fishing, and marine pollution.

Regimes treated as collections of rights, rules, principles, decision-making procedures, and social practices differ from organizations treated as material entities that have offices, personnel, budgets, and legal personality (Young 1989a). The point of this distinction is not to suggest that regimes (or institutions more generally) are more important than organizations or vice versa. Rather, the distinction allows us to think about relations between regimes and organizations as a topic for systematic investigation. Do regimes always require organizations to implement or administer their provisions under real-world conditions or to ensure that subjects comply with their terms? Can organizations operate independently of regimes or even take the initiative in creating new regimes to address a range of substantive concerns?

The effectiveness of environmental and resource regimes is a function of the extent to which these arrangements contribute to solving or mitigating the problems that lead to their creation (Young 2011a). Analyses of the effectiveness of regimes often include distinctions among outputs, outcomes, and impacts. Outputs involve the promulgation of regulations and the establishment of infrastructure needed to move a regime from paper to practice. Outcomes refer to changes in the behavior of those subject to the rights, rules, principles, and decision-making procedures of a regime. Impacts are a matter of problem-solving as such. There is no way for a regime to solve a problem without producing outputs and outcomes. But the production of outputs and outcomes provides no guarantee that a regime will prove successful in solving problems. I treat outputs and outcomes as important topics to consider but always in relation to problem-solving. Other criteria of evaluation are useful in assessing the performance of a regime, including sustainability, efficiency, equity, and various considerations of good governance (Young 2013b). It is important to solve problems, but it is desirable to do so in a manner that is efficient, equitable, and democratic as well.[5]

General Observations about Environmental Governance

With these definitions in hand, we are ready to move on to an examination of insights arising from the study of environmental problem-solving that seem relevant to the challenge of governing complex systems. I start with ten observations that are general in nature in the sense that they apply to environmental governance in all settings. These generalizations are not products of some rigorous effort at hypothesis testing. Rather, they take the form of insights that are derived from my long-standing effort to understand the nature and workings of environmental governance systems and that I have found particularly robust.

Since no one can keep ten observations in mind at the same time, however, let me summarize the take-home message at the outset. Environmental and resource regimes play major roles in determining the extent to which societies are able to solve a wide range of problems arising in human-environment relations. But a cookie-cutter approach to the formation and implementation of regimes cannot succeed. Regimes interact with a variety of biophysical and socioeconomic forces that affect their performance in specific settings. Successful regimes are those that are well-matched to the principal features of the biophysical and socioeconomic settings in which they operate.

Observation 1: Governance without government is a common phenomenon. The existence of a government in the ordinary sense of the term is certainly not sufficient to assure success in fulfilling the function of governance. But my emphasis here is on the observation that the existence of a government also is not necessary. The significance of this proposition is that it allows us to separate the study of governance treated as a social function from research focusing on the nature and performance of governments. There are many cases in which governments perform poorly in fulfilling the function of governance. Where governments come under the sway of dictators or fall prey to corruption, they may do more to block or undermine efforts to perform the function of governance than to promote progressive results. There is nothing surprising about this observation. More interesting in terms of the production

Table 1.1
General observations about environmental governance

Observation 1: Governance without government is a common phenomenon.

Observation 2: Three distinct processes dominate both the formation of environmental and resource regimes and their evolution over time: spontaneity/self-generation, negotiation/bargaining, and coercion/imposition.

Observation 3: Institutional bargaining focuses on building coalitions of those willing to contribute to the supply of public goods rather than on reaching agreement on mutually acceptable outcomes located at specific points on contract curves.

Observation 4: Although it is common to think of regimes as regulatory arrangements, they can and often do perform procedural, programmatic, and generative functions.

Observation 5: Individuals play important roles both in the formation of environmental and resource regimes and in efforts to maximize their effectiveness once they are in place.

Observation 6: Environmental and resource regimes can guide the behavior of actors either by alleviating collective-action problems or by encouraging the development of social practices.

Observation 7: Effective and resilient environmental and resource regimes generally rest—implicitly if not explicitly—on the cognitive foundation provided by a prevailing discourse or worldview.

Observation 8: High levels of uncertainty alter normal utilitarian calculations guiding the formation and operation of environmental and resource regimes.

Observation 9: The success of environmental and resource regimes is a function of the fit or match between the principal elements of these institutional arrangements and the major features of the biophysical and socioeconomic settings in which they operate.

Observation 10: Although some propositions about the effectiveness of environmental and resource regimes scale up/down in the dimension of social organization, others do not.

of new knowledge is the realization that there are situations in which it is possible to perform the function of governance in the absence of a government (Rosenau and Czempiel 1992). This realization has triggered a growing body of literature on mechanisms that yield favorable results in situations in which there is no government in the ordinary sense of the term. Although this observation applies across a range of issue areas, it has proven particularly valuable for those of us concerned with human-environment relations. Small-scale stateless societies often succeed in avoiding or mitigating problems like the "tragedy of the commons" through the development of regimes that prove successful in the absence of anything recognizable as a government (Hardin 1968; Ostrom et al. 2002). Especially important now, as we endeavor to solve large-scale problems like climate change and the loss of biological diversity that have emerged as great issues of our time, is the observation that the absence of anything resembling a supranational government does not condemn us to a future in which threats to individual well-being and social welfare become ever more severe (Young, King, and Schroeder 2008). This line of thinking has led also to a lively discussion about the extent to which propositions about the determinants of effectiveness in environmental and resource regimes scale up/down across levels of social organization (Dietz, Ostrom, and Stern 2003; Young 2005a).

Observation 2: Three distinct processes dominate both the formation of environmental and resource regimes and their evolution over time: spontaneity/self-generation, negotiation/bargaining, and coercion/imposition. Like all social institutions, regimes may simply develop as a matter of practice in the absence of intentional initiatives on the part of anyone, much less the development of a rule book to guide the actions of those who participate in these arrangements. Such arrangements are known as spontaneous or self-generating regimes (Hayek 1973). On the other hand, bargaining about the attributes of institutions is a common occurrence at all levels of social organization, and it is clearly the case that powerful actors often exercise disproportionate influence when it comes to working out the defining features of environmental and resource regimes. This is particularly apparent at the international level, where efforts to negotiate the terms of multilateral environmental agreements constitute a prominent feature of the political landscape, and dominant actors—often referred to as hegemons—can play exceptionally influential roles in the creation of regimes treated as public goods. Needless to say, these distinctions are analytic in nature (Young 1982a; Young 1982b). Actual cases typically involve elements of two or even all three of these processes. One or a few powerful actors regularly wield exceptional influence, for example, when it comes to hammering out key provisions of the conventions or treaties setting forth the terms of international regimes. Under the circumstances, the interesting challenges involve identifying scope conditions that govern the roles that each of the three processes play in real-world situations and understanding interactions among the three processes as determinants of the principal attributes of the regimes that emerge to address specific problems arising in human–environment relations.

Observation 3: Institutional bargaining focuses on building coalitions of those willing to contribute to the supply of public goods rather than on reaching agreement on mutually acceptable outcomes located at specific points on contract curves. Institutions are public goods in the sense that once supplied to a group, they are largely non-excludable as well as non-rival for the individual members of the relevant group (Young 1989b). Individual members may have strong preferences among available institutional options; they may even regard some options as public "bads." They will naturally campaign vigorously for their preferred options. But this does not alter the fact that institutions are largely non-excludable and non-rival.[6] This feature of environmental and resource regimes (or any other institutions) has profound implications for the character of the negotiations or bargaining processes that often play a central role in regime formation. In conventional bargaining, the participants have incentives to form minimum winning coalitions to arrive at agreements that reduce the number of actors expecting to share in the proceeds (Riker 1962). But when the proceeds take the form of public goods, there are no such incentives. Non-rivalness ensures that enlarging the size of the group does not diminish the benefits accruing to individual members. In institutional bargaining, the focus is apt to shift to ensuring that participants implement and live up to the obligations they assume, and it is widely expected that those who have a voice in crafting the terms of regimes will be more likely to regard them as legitimate and, as a result, to live up to the obligations they assume on a voluntary basis. To the extent that this is the case, those engaged in institutional bargaining have an incentive to create maximum winning coalitions in contrast to minimum winning coalitions. The attractions of this strategy will be particularly strong in settings like international society where the fulfillment of obligations on an ongoing basis is to a large degree voluntary (Young 1994a).

Observation 4: Although it is common to think of regimes as regulatory arrangements, they can and often do perform procedural, programmatic, and generative functions. There is a natural tendency to equate institutions in general with the rules of the game in the domains in which they operate. Certainly, this is a key feature of many environmental and resource regimes. Think of the rule requiring emitters of sulfur dioxide to have permits in the US clean-air regime or the rule prohibiting mining activities in the Antarctic Treaty System. But regulation is not the only function that regimes perform (Young 1999). They often handle procedural tasks, like setting annual quotas in commercial fisheries or making decisions about phaseout schedules for ozone-depleting substances. Some regimes focus on programmatic activities, such as the actions needed to clean up the Rhine River or the remedial action plans developed to improve water quality in the Great Lakes of North America. Still others play generative roles in enhancing knowledge about the relevant problems (e.g., the operation of the assessment program in the European long-range transboundary air pollution regime), making judgments about the extent and severity of the problems, framing them for inclusion on policy agendas, and pushing them toward the top of the agenda in relevant policy forums. These functions are not mutually exclusive; the same regime may be structured to address two or more of them. One or another of these functions may take on particular importance at various stages in the life of a single regime.

Observation 5: Individuals play important roles both in the formation of environmental and resource regimes and in efforts to maximize their effectiveness once they are in place. There is a tendency among some observers, particularly those who think about international politics and focus on the roles that nation states play, to downplay the influence of the actions of individuals in creating and implementing regimes. But this turns out to be a mistake. Of course, it is essential in thinking about leadership on the part of individuals to avoid the endogeneity trap. If we use some measure of their contributions to identify those who are leaders, we can no longer ask questions about the extent to which the activities of leaders make a difference. But it is fairly easy to avoid this trap by identifying a universe of cases populated by those who have occupied particular roles (e.g., chief negotiators, chairs of negotiating committees), and then asking questions about their influence with respect to the creation and operation of regimes. Not all members of this universe prove to be influential. But it is striking how often such individuals do make a difference in cases characterized by success in forming regimes and by effectiveness in the operation of regimes (Young 1991; Young and Osherenko 1993; Young

1998). It turns out that there are several distinct types of leadership that are worthy of separate treatment: (i) cognitive leadership or the ability to come up with new ways of thinking about key issues, (ii) entrepreneurial leadership or the ability to exercise skill in making deals or devising the terms of mutually acceptable agreements, and (iii) structural leadership or the ability to bring the influence of powerful actors to bear on specific issues in a constructive manner. These types of leadership are not mutually exclusive; the same individual may excel in more than one domain of leadership. Nevertheless, the activities involved are so different that the ability to operate successfully in more than one mode is exceptional.[7]

Observation 6: Environmental and resource regimes can guide the behavior of actors either by alleviating collective-action problems or by encouraging the development of social practices. Two distinct analytic perspectives dominate thinking about the mechanisms through which regimes operate to steer the behavior of the actors engaged in human–environment relations.[8] Those who think about such matters in broadly utilitarian terms and focus on the behavior of self-interested actors look to regimes to solve or alleviate collective-action problems by reassuring individual actors regarding the cooperative behavior of others, linking the reputations of actors to compliant behavior, and lengthening the shadow of the future. Those who adopt a social-practice perspective, by contrast, are more likely to link the influence of regimes to mechanisms like encouraging socialization, shaping the content of social norms or principles, and embedding compliance in standard operating procedures. The two sets of mechanisms are not mutually exclusive. The same regime may contribute to the effectiveness of a governance system by influencing calculations of the costs and benefits of (non)compliance, and by inculcating a willingness on the part of actors to comply with prescriptions as a matter of habit or without engaging in an effort to make deliberate calculations regarding costs and benefits. A general theory of environmental governance ought to integrate these perspectives into a comprehensive account of the influence of institutions on the course of human–environment relations.

Observation 7: Effective and resilient environmental and resource regimes generally rest—implicitly if not explicitly—on the cognitive foundation provided by a prevailing discourse or worldview. Most environmental and resource regimes are focused arrangements consisting of collections of rights, rules, decision-making procedures, and practices applicable to specific situations. Fisheries regimes operating in domestic settings feature procedures for setting allowable harvest levels on an annual basis, allocating quotas among fishers, devising gear restrictions, and so forth. At the international level, the ozone regime covers several families of chemicals and specifies phaseout schedules for each of the regulated chemicals. Yet these specific arrangements reflect, implicitly if not explicitly, broader and deeper presumptions about proper ways of organizing human activities in the relevant domains. A familiar example of this phenomenon is associated with the idea of "embedded liberalism" in the realm of economic affairs (Ruggie 1982). The concrete elements of the international monetary and trade regimes are based on a set of propositions calling for a clear separation between the public sector and the private sector, coupled with an understanding that it is appropriate for governments to intervene from time to time by deploying monetary and fiscal policies or supporting infant industries or industries deemed essential to national security. Parallel, though perhaps less familiar, ideas dealing with human-environment relations feature propositions about the role of free enterprise coupled with regulatory measures needed to avoid the tragedy of the commons and to deal with environmental side effects or externalities, while pursuing maximum sustainable yields. The point of this proposition is not to extol the virtues of any particular discourse or worldview. It merely asserts that a link to some widely shared cognitive underpinnings is an important condition for the achievement of effectiveness in the operation of regimes dealing with specific issues.

Observation 8: High levels of uncertainty alter normal utilitarian calculations guiding the formation and operation of environmental and resource regimes. Real-world conditions often feature high levels of uncertainty both about the costs of failing to take action by (re)forming suitable regimes and about the costs and benefits of creating such regimes and operating them effectively. The case of climate change provides a prominent example. Opinions vary dramatically about the likely consequences for social welfare of a failure to control climate change, even over the short run. There are profound differences in estimates of the probable costs of taking effective steps to address the problem of climate change in a timely manner. Such situations give rise to a "veil of uncertainty," ensuring that straightforward calculations of benefits and costs will not

suffice as a basis for making sensible decisions about the (re)formation
of environmental regimes (Brennan and Buchanan 1984; Young 1994a).
The thicker the veil, the more likely actors are to respond by imposing
conditions on the applicability of their commitments, entering into agreements that are easy to opt out of or rescind, or establishing sunset provisions that limit the length of any commitments they make. High levels of
uncertainty also may increase the receptivity of those engaged in regime
formation to considerations of fairness or justice. To the extent that it is
difficult to forecast the incidence of the costs of inaction as well as the
costs of taking concrete steps to solve specific problems, the parties to
regimes will have incentives to favor arrangements that produce results
regarded as equitable for all those concerned.

*Observation 9: The success of environmental and resource regimes is
a function of the fit or match between the principal elements of these
institutional arrangements and the major features of the biophysical and
socioeconomic settings in which they operate.* Some scholars have argued
that it is possible and helpful to rank environmental problems on a scale
ranging from easy or benign problems at one end to difficult or malign
problems on the other end (Miles et al. 2002). On this account, problems
like climate change are extremely malign—some even call them wicked
or diabolical—and we should not expect to be able to solve them during the foreseeable future (Steffen 2011). My view, by contrast, is that
the critical challenge is to pinpoint the essential features of problems
and then to construct regimes that are well matched to the problems at
hand in the sense that they are crafted in such a way as to be responsive
to these essential features (Young 2002a; Galaz et al. 2008). There are
cases in which this approach has yielded successful solutions to seemingly
malign problems. The Antarctic Treaty of 1959, for instance, emerged in
the midst of the Cold War, involved active cooperation on the part of the
superpowers, and included arrangements dealing with notoriously challenging problems like demilitarization and disputes over jurisdictional
claims. Conversely, efforts to address seemingly benign problems can
fail when those who design regimes come up with arrangements that are
poorly matched to the problems at hand. Many regional fisheries regimes,
for example, fail or perform poorly because they do not include effective
procedures for setting allowable harvest levels or allocating catch shares
among harvesters, even when their members are generally friendly and

have cooperated successfully on other issues. The principal insight to be drawn from these observations is that we need to strengthen our skills in the realm of what I (and others) call institutional diagnostics (Young 2002a; Ostrom et al. 2007; Young 2008). The road to success in devising effective environmental and resource regimes lies in diagnosing the essential features of specific problems and then devising institutional arrangements that are crafted carefully to address the key issues.

Observation 10: Although some propositions about the effectiveness of environmental and resource regimes scale up/down in the dimension of level of social organization, others do not. The issue of scale is a prominent concern among ecologists who think regularly about the generalizability of their findings across levels ranging from leaf to landscape. This concern is less familiar to social scientists thinking about generalizability across levels of social organization, from the micro-level of local systems to the macro-level of global systems. Yet it is natural to pose such questions with regard to the effectiveness of governance systems (Young 1994b; Young 2002a; Dietz, Ostrom, and Stern 2003). Do findings about the conditions under which small-scale societies succeed in avoiding or mitigating the tragedy of the commons apply to thinking about ways to protect global commons like the Earth's climate system? Do ideas emanating from analyses of international environmental governance (e.g., the concept of regime complexes) have counterparts in thinking about the governance of local commons? There are particularly interesting parallels in this regard between the micro-level and the macro-level, since many small-scale societies are stateless in a manner that resembles the statelessness of international society. In both cases, the issue of providing governance without government becomes a central concern (Young 2005a). But a more systematic assessment regarding prospects for scaling up/down must delve into similarities and differences relating to the behavior of the actors involved, the characteristics of the issues they seek to address, the nature of the institutional context, and key features of the broader setting (e.g., the nature of prevailing technologies). It is easy to see that facile efforts to scale up/down across levels of social organization will almost always fail. Nevertheless, there is much to be learned about the factors that determine the effectiveness of environmental and resource regimes by comparing and contrasting the insights about such matters produced by analysts who are steeped in the forces at work in different social settings.

Insights Specific to International Environmental Governance

I was trained originally as a political scientist specializing in the subfield of international relations. I began to take a professional interest in environmental issues during the run up to the 1972 UN Conference on the Human Environment, and I developed an interest in international environmental governance at an early stage. The bulk of my applied work through the years has dealt with efforts to solve large-scale problems relating to the polar regions, the oceans, and the Earth's atmosphere. It is not surprising, then, that some of my enduring insights focus on international environmental governance treated as a proper subset of the overarching category of environmental governance. In this section, I present five observations emerging from my work on environmental governance that are applicable to regimes operating in international society, though they do not generalize to other levels of social organization.

Insight 1: The anarchic character of international society complicates but does not rule out efforts to create effective environmental and resource regimes. Perhaps the first issue to tackle in this realm arises from the fact that international society lacks a government in the ordinary sense and that this condition is widely regarded as a serious limitation

Table 1.2
Insights specific to international environmental governance

Insight 1: The anarchic character of international society complicates but does not rule out efforts to create effective environmental and resource regimes.
Insight 2: International regimes interact with one another in a variety of ways that have significant consequences for their effectiveness.
Insight 3: While the subjects of international regimes are normally states, the success of these arrangements depends on both the capacity and the willingness of member states to implement their requirements in domestic political arenas.
Insight 4: Institutional dynamics affecting international regimes produce a diverse but limited set of outcomes best understood as emergent properties of complex systems.
Insight 5: The onset of the Anthropocene puts a premium on the creation and operation of international environmental and resource regimes that are effective in turbulent times and capable of adapting nimbly or agilely to rapidly changing conditions.

when it comes to meeting needs for governance at the international level (Bull 1977). There is even a strand of thinking leading to the conclusion that the underdeveloped character of enforcement mechanisms in international society makes the idea of international law an oxymoron. But these views are exaggerated. If we start by separating compliance, which is a matter of behavior that conforms to the requirements of prescriptions, from enforcement, which is a matter of applying sanctions to elicit compliance, it becomes clear that the absence of effective enforcement mechanisms is not a fatal flaw (Young 1979). A number of factors are relevant in this connection (Young 1999). Some regime functions (e.g., capacity building, the production of scientific assessments) are not regulatory in nature and do not give rise to issues of compliance. Some regulations (e.g., those adopted to address what are known as coordination problems) do not generate incentives to cheat. Some regulatory systems can tolerate a significant amount of non-compliance without collapsing or becoming ineffectual. Some non-compliance procedures (e.g., those developed under the terms of the Montreal Protocol on stratospheric ozone depletion) address issues of non-compliance without resorting to the use of sanctions in the ordinary sense. Some regimes acquire domestic constituencies supporting compliance either in the form of lead agencies responsible for their implementation or in the form of sizable groups of supporters in the general public (Dai 2005; Dai 2007). The core elements of powerful discourses can become normative precepts that influence the behavior of decision makers, whether or not they are backed by sanctions. None of this leads to the conclusion that there is no need to worry about compliance with the terms of international regimes. But it would be a serious mistake to assume that these arrangements are bound to fail simply because there is no government in the ordinary sense to enforce the rules of international governance systems.

Insight 2: International regimes interact with one another in a variety of ways that have significant consequences for their effectiveness. Distinct regimes operating in the same broad issue domain often interact with one another. The interplay between the ozone regime and the climate regime constitutes a prominent example.[9] Environmental regimes also interact with regimes operating in other issue areas. The interplay between the ozone regime and the trade regime exemplifies this type of interaction. In my early work on this phenomenon, I distinguished among (i) embedded regimes or arrangements rooted in some deeper social order or common worldview, (ii) nested regimes or arrangements dealing with distinct issues but based on a common constitutive framework, (iii) clustered regimes or non-hierarchical arrangements linked to one another in spatial or functional terms, and (iv) overlapping regimes or separate arrangements that interact in unintended and often unforeseen ways (Young 1996a). Others have picked up on this general theme and introduced a variety of important distinctions going beyond my early formulation. They have been particularly successful in shedding light on the consequences arising in situations featuring overlapping regimes (Oberthür and Gehring 2006; Oberthür and Stokke 2011). In my judgment, there is much to be learned about the roles that nested and clustered regimes play in efforts to solve complex environmental problems. The analysis of nested regimes directs attention to the links between constitutive arrangements (e.g., the UN Convention on the Law of the Sea) and a variety of more operational arrangements built on a common constitutive base (e.g., regimes dealing with commercial shipping, industrial fisheries, and marine pollution). The idea of clustered regimes now appears as a precursor to the productive analysis of regime complexes that has flourished in recent years (Raustiala and Victor 2004; Keohane and Victor 2011; Orsini, Morin, and Young 2013). Such complexes, dealing with multidimensional issues, like Antarctica, plant genetic resources, or climate change, offer promising ways forward in situations where it is politically infeasible to devise fully integrated governance systems covering all aspects of a spatially or functionally delimited topic or theme.

Insight 3: While the subjects of international regimes are normally states, the success of these arrangements depends on both the capacity and the willingness of member states to implement their requirements in domestic political arenas. International regimes differ from governance systems operating in other settings both because there is no government in the ordinary sense to induce or compel states to comply with their commitments, and because states may find it difficult to ensure compliance with such commitments on the part of domestic actors even when their intentions are honorable. The issues arising from the anarchic character of international society are familiar to all those who work in this field. But there is much more to be said about the treatment of international commitments in domestic settings (Putnam 1998; Dai 2005; Dai 2007).

Ratification typically makes international commitments legally binding on actors operating in domestic settings. But other mechanisms may play equally important roles in turning international commitments into legally and politically potent forces affecting the behavior of member states and their domestic constituents. These include: (i) designating a lead agency for purposes of implementation that subsequently becomes a staunch advocate for the regime in domestic politics (e.g., the US Department of the Interior in the case of wildlife regimes), (ii) embedding the provisions of a regime in thick social practices even in the absence of ratification (e.g., the American practice of adhering to most of the provisions of UNCLOS), and (iii) using the existence of the regime as an important argument in domestic legal proceedings (e.g., reliance on the terms of international agreements relating to migratory birds in domestic litigation addressing the scope of the authority of the Federal government in the United States). None of this alters the challenge of making international regimes effective in a world in which states are sovereign with regard to their roles in international society but frequently weak with regard to their ability to make domestic constituents comply with their international commitments. But these factors do go some way toward explaining why the anarchical character of international society is not a fatal flaw when it comes to the creation of successful environmental regimes.

Insight 4: Institutional dynamics affecting international regimes produce a diverse but limited set of outcomes best understood as emergent properties of complex systems. Environmental and resource regimes are highly dynamic. Once formed and put into practice initially, they experience continual change. Both internal or endogenous forces and external or exogenous forces play major roles in influencing the trajectories of institutional change over time; the outcomes typically reflect the interplay of internal and external forces (Young 2010). We can identify a number of distinct patterns in the emergent properties arising from such processes. International regimes may experience: (i) progressive development in the sense of moving from strength to strength toward success in problem-solving, (ii) punctuated equilibrium in the sense of experiencing relatively long periods of stasis separated by shorter bursts of change, (iii) arrested development in the sense of running into severe obstacles or road blocks that become lasting barriers to success, (iv) diversion in the sense of a more or less sharp deviation from the goals or purposes agreed to at the outset, and (v) collapse in the sense of disintegration and disappearance in the wake of changed circumstances. Which of these patterns occurs in specific cases is determined by the alignment and interaction of internal processes and external forces. A regime that is relatively simple and lacking in capacity to adapt to changing circumstances, for example, may perform perfectly well in a stable setting in which the identity of the actors and the character of the problem remain unchanged. But such a regime may simply collapse and be swept away in the face of changes in the biophysical or socioeconomic setting that it has no capacity to control. A case in point is the North Pacific fur seal regime. Students of wildlife management long regarded this regime, launched initially in 1911 to cope with severe depletions of fur seal stocks resulting from commercial harvesting, as a striking case of success. But the regime collapsed almost overnight in 1984–1985, following dramatic changes in the jurisdiction of coastal states over marine systems and sharp shifts in the condition of the Bering Sea ecosystem (Young 2005a). As this discussion suggests, the analysis of institutional dynamics constitutes a cutting-edge concern regarding our understanding of international regimes.

Insight 5: The onset of the Anthropocene puts a premium on the creation and operation of international environmental and resource regimes that are effective in turbulent times and capable of adapting nimbly or agilely to rapidly changing conditions. The Earth is moving out of the Holocene, a period of roughly 10,000 years marked by relative stability and conditions that were benign from the perspective of humans, and into the Anthropocene, a period of human-dominated systems prone to changes that are far-reaching and sometimes abrupt but difficult to anticipate (Steffen et al. 2004; Steffen, Grinevald, Crutzen, and MacNeill 2011). The requirements for success in an era marked by changes that are nonlinear, sometimes abrupt, typically irreversible, and often nasty from the perspective of human welfare differ significantly from those characteristic of more settled times (Young 2012a; Young 2012b). From the point of view of effectiveness, these conditions place a premium on the ability to anticipate state changes in both biophysical and socioeconomic systems, to take steps to avoid passing tipping points, and to adapt quickly to changing circumstances on an ongoing basis. The point of this observation is not to cast doubt on our ability to solve the large-scale governance problems coming into focus as we move deeper into the Anthropocene.

Rather, the take-home message is that standard operating procedures will need to be reexamined and in some cases reformed if we are to achieve success in solving problems under these new conditions. We will need to devote more resources to the operation of effective early warning systems, develop procedures that enhance decision-making under uncertainty, and move away from the delays and even paralysis that often accompany reliance on consensus procedures. There is no reason to conclude that we will be unable to create international regimes that produce successful outcomes under the conditions characteristic of the Anthropocene. But the social capital needed for success in this new era is likely to differ substantially from the social capital we have relied on in the past (Brondisio, Ostrom, and Young 2009).

A Concluding Thought

This chapter sets the stage for the analysis to come. The conceptual framework presented in the first substantive section is generic. It can be used to frame questions and report findings about efforts to fulfill needs for governance in any setting. The insights reported in the second and third sections, by contrast, are subject to critical assessment as we move from one setting to another. It is appropriate to ask how well these findings will hold up as we seek to address emerging needs for governance in complex systems. Before taking up this challenge, however, I turn in the next chapter to a more detailed account of what we mean in talking about the effectiveness of governance systems and what we know about the determinants of effectiveness. In the final analysis, we are interested in analyzing governance systems because we think they are significant drivers of the trajectory of human affairs in general and, more specifically of the course of human-environment interactions.

2

Determinants of Effective Problem-Solving

Introduction

A common observation among those concerned with solving environmental problems and, more generally, promoting sustainability in human-environment relations is that governance systems work relatively well at the national level but poorly or not at all in efforts to solve international, transnational, and especially global problems of the sort that have become a growing focus of attention in the Anthropocene (Speth 2004).[1] While the state is a positive force in steering collective behavior regarding the management of natural resources and the protection of the environment in domestic settings, this argument runs, the anarchic character of international society treated as a society of spatially delimited sovereign states constitutes a barrier to successful governance at the international level.

Yet both elements of this line of thinking are open to question. Failures to tackle environmental problems effectively, much less to achieve sustainability in human-environment relations, are common not only in developing societies facing severe problems of poverty and hunger or saddled with the curse of natural resources but also in advanced industrial societies (Collier 2008). Although efforts to address the global challenges of climate change, the loss of biological diversity, and the degradation of ecosystem services leave a great deal to be desired, international environmental governance does not present a uniform picture of failure.

Some international environmental governance systems or, as they are commonly called, international regimes make a difference in the sense that they contribute to solving large-scale problems. Arrangements widely regarded as effective in these terms include the global regime created to protect the stratospheric ozone layer, the international governance system applicable to Antarctica, and the regional arrangement established to clean up the Rhine River.[2] The lack of efficacy or relative failure of other regimes created to deal with environmental problems is equally evident. Prominent examples include the climate regime, the arrangement created to combat desertification, and some (but not all) of the regional fisheries management arrangements.[3] Many regimes fall between these polar categories. They achieve a measure of effectiveness, though it is often hard to place them precisely along a continuum ranging from total failure to clear-cut success. Cases that fit this description include the regime dealing with pollution of the sea from ships, the arrangements addressing pollution in the North Sea, the regime governing trade in endangered species, and the regime articulated in the Great Lakes Water Quality Agreement. Not surprisingly, some regimes are successful for a time but subsequently decline or even collapse (e.g., the regime for North Pacific fur seals), while others are slow to gain traction but become more effective with the passage of time (e.g., the transboundary air pollution regime in Europe). In still other cases (e.g., the climate regime), it is difficult to forecast whether or not a regime will reach a threshold and gain traction over the course of time.

In this chapter, I focus intensively on the determinants of effectiveness in an effort to account for this mixed record in efforts to solve problems involving human-environment relations in international society. Picking up on themes introduced in a general way in the previous chapter, I drill down on particular factors thought to account for variations in levels of effectiveness in order to answer the following questions. Can we formulate well-defined and well-supported propositions about the determinants of success and failure? Are we able to arrive at conclusions that will be of interest both to those responsible for implementing the provisions of existing environmental regimes, and to those engaged in efforts either to strengthen existing arrangements or to create entirely new ones?

I address these questions in three steps. The first step involves conceptual and definitional issues; it focuses on clarifying the meaning of effectiveness with regard to regimes dealing with human-environment relations. The second step centers on identifying and discussing the most important things we have learned so far about the determinants of institutional effectiveness. Step three features an exploration of cutting-edge themes or areas ripe for increased attention on the part of researchers going forward. The take-home message is one of cautious optimism. There is much we can do to add to our understanding of ways to meet the challenge of solving problems involving human-environment relations, despite the impacts of some obvious as well as some less apparent limitations on the methods available for pursuing this goal.

What Do We Mean by Effectiveness?

The concept of effectiveness as applied to environmental regimes is complex and subject to a variety of formulations (Mitchell 2008; Underdal 2008). Arguably, the core concern is the extent to which regimes contribute to solving or mitigating the problems that motivate efforts to create them (Young 1999a). But there are other ways to think about effectiveness that are both less ambitious and more ambitious than this focus on problem-solving. Less ambitious conceptions of effectiveness direct attention to what are known as (i) outputs or the regulations and administrative arrangements created to move a regime from paper to practice and (ii) outcomes or changes in the behavior of actors relevant to the problem at hand (Young 1999b). Success in these terms is significant, but it does not guarantee progress in solving the relevant problems. More ambitious conceptions, on the other hand, seek to assess the performance of regimes relative not only to the probable course of events in their absence (i.e., the no-regime counterfactual) but also to some conception of an ideal outcome often called the collective optimum. The effectiveness of a regime (E) is then measurable as the location of actual performance (AP) on the spectrum ranging between the no-regime counterfactual (NR) and the collective optimum (CO) or:

$$E = \frac{AP\text{-}NR}{CO\text{-}NR} \tag{1}$$

Normalizing this equation by setting NR equal to 0 and CO equal to 1 produces a way to compare and contrast the effectiveness of different regimes on a common scale. This formulation is conceptually appealing. Yet it is hard to operationalize the terms of this equation. The meaning of CO is often unclear, and we are seldom able to arrive at any straightforward calculation of the value of NR (Helm and Sprinz 2000; Hovi, Sprinz, and Underdal 2003a; Young 2003; Hovi, Sprinz, and Underdal 2003b).

Several additional aspects of effectiveness deserve notice at the outset. A regime's participants may differ both in the importance they attach to the problem and in the way they frame it for consideration in policy forums. Those who create regimes may harbor unstated goals that differ significantly from those spelled out in constitutive documents. The effectiveness of regimes may vary through time. Some regimes go from strength to strength with the passage of time. Others are relatively ineffective at the outset but gain strength over time or vice versa. Many of those seeking to assess the effectiveness of regimes add other measures of success to the core concern of problem-solving, including some criterion of institutional resilience, a concern for economic efficiency, various measures of fairness or equity, and one or more considerations embedded in the idea of "good" governance (Mitchell 2008).

Evaluating the effectiveness of environmental regimes is a challenging task under the best of circumstances (Underdal 2008). In every case, we want to compare the actual course of events regarding the relevant problem with what would have happened in the no-regime counterfactual. Although this is easier to do with regard to some measures of effectiveness (e.g., outputs) than others (e.g., problem-solving), documenting the consequences resulting from the creation and operation of a regime is always demanding. Additionally, regimes invariably operate in complex settings in which a variety of other forces are at work. Separating the signal attributable to the operation of a regime from the noise associated with the influence of a variety of other forces at work at the same time is a difficult task. Those endeavoring to address these issues have employed a number of methodological tools, singly or in combination (Underdal and Young 2004; Young et al. 2006). Suffice it to say here that some differences of opinion regarding the effectiveness of regimes are more apparent than real in the sense that they are artifacts of the definitions of effectiveness selected or the procedures used to evaluate effectiveness, rather than substantive disagreements about the actual performance of specific regimes.

What Do We Know about Institutional Effectiveness?

Scientists understandably focus on cutting-edge questions that constitute the frontiers of research in their areas of interest, a practice that directs attention to issues that we do not understand or at least do not understand well. But in this discussion of the current state of knowledge regarding the effectiveness of regimes dealing with human-environment relations, it is appropriate to begin with an account of what we have learned so far about this subject. I address this topic under three headings: general findings about effectiveness, findings about specific determinants of success, and findings about institutional interplay.

Three distinct bodies of evidence deserve attention in this assessment: qualitative case studies carried out most often by analysts trained as political scientists (Haas, Keohane, and Levy 1993; Skjaerseth 2000; Wettestad 2002; Parson 2003), quantitative case studies typically produced by analysts with a background in economics (Murdoch and Sandler 1997; Murdoch, Sandler, and Sargent 1997; Finus and Tjøtta 2003; Ringquist and Kostadinova 2005), and quantitative analyses that seek to develop generalizations about effectiveness drawing on evidence from sizable universes of cases (Miles et al. 2002; Breitmeier, Young, and Zürn 2006; Breitmeier, Underdal, and Young 2011). The conclusions emerging from these bodies of evidence overlap. But they are not entirely compatible. Those who have carried out the qualitative case studies, perhaps reflecting a positive attitude toward the role of institutions common among political scientists, tend to find evidence of the significance of regimes in addressing environmental problems. The quantitative case studies, arguably reflecting skeptical attitudes toward governance systems common among economists, typically raise doubts about the roles regimes play. The large-N studies have sought to move beyond this divide, endeavoring to discriminate among cases in which regimes matter a lot or a little and seeking to identify the determinants of success and failure.

General Findings

Though it may be a source of frustration to those hoping for simple generalizations regarding the determinants of effectiveness, differences in the findings flowing from the three bodies of evidence are understandable. In virtually every case, a regime constitutes only one of a number of distinct forces influencing the course of human-environment relations. In most cases, these forces are highly interactive (Young 2002b). Still, there are some general findings about the effectiveness of environmental regimes arising from the research carried out so far. In this subsection, I comment on what seem to me to be the most important of these findings

Some regimes matter in the sense that they make a (sometimes sizable) *difference not only in terms of outputs and outcomes but also in terms* *of solving the problems that motivate their creation.* It is easy to overestimate the success of environmental regimes. Quantitative case studies, rooted in a rational-choice paradigm, have suggested that key actors may have reduced the production of ozone-depleting substances or emissions of airborne pollutants voluntarily, that non-regime factors may account for as much or more of the success in dealing with water pollution as the operation of the relevant regimes, and that actual outcomes fall short of the collective optimum in most cases. Yet in-depth qualitative case studies, making extensive use of procedures like process tracing and thick description, have concluded that regimes have contributed to the development of new knowledge and new social practices that have played important roles in dealing with long-range transboundary air pollution in Europe (Wettestad 2002), the depletion of the stratospheric ozone layer (Parson 2003), the control of pollution in the North Sea (Skjaerseth 2000), and the management of commercial fisheries in the Barents Sea (Stokke 2012). In an effort to reconcile these findings, several teams of researchers have created databases containing sufficiently large numbers of cases to allow for the development of empirical generalizations about the effectiveness of environmental regimes. Miles et al., drawing on a dataset including 37 cases, report that 50 percent of these regimes produced behavioral changes and 35 percent played a significant role in terms of problem-solving (Underdal 2008: 59). Breitmeier, Young, and Zürn, employing a dataset encompassing 172 cases, report that in situations where problems improved slightly or considerably, regimes had a "significant" or "very strong" influence 52 percent of the time (Underdal 2008: 59). Environmental regimes can make a difference. But they do not always work, and they never operate in a vacuum devoid of other causal forces (Breitmeier, Underdal, and Young 2011).

The anarchic character of international society is not always an *obstacle to creating and implementing effective regimes.* Many observers regard the absence of a government at the international level as a severe impediment to the establishment of effective regimes, primarily because it rules out the use of enforcement mechanisms of the sort that states employ to induce their subjects to comply with the provisions of domestic regulatory systems. While this is certainly a concern in some cases, it does not loom large in situations where compliance on the part of most members of the group is unnecessary, the parties to environmental agreements have no incentive to cheat, factors other than sanctions in the ordinary sense provide subjects with good reasons to comply, or various forms of private or hybrid governance are able to exert pressure on subjects to comply (Young 1999a; Barrett 2007; Dai 2007). There is no basis for complacency here when it comes to dealing with the great issues of our times, such as climate change, the loss of biological diversity, and, more generally, sustainable human-environment relations. But neither is there any basis for dismissing the capacity of regimes to contribute to solving a range of problems.

Regime design is often a more significant determinant of effectiveness than some measure of whether the problem is benign (i.e., easy to solve) or malign (i.e., hard to solve). Poorly designed regimes can produce disappointing results even in cases where problems are straightforward and relatively easy to solve; well-designed regimes can produce positive results even in dealing with problems that are widely regarded as malign. This has given rise to a stream of research on what has become known as the issue of fit (Galaz et al. 2008), together with a growing interest in institutional diagnostics (Young 2002a; Ostrom et al. 2007; Young 2008). Whereas the effort to conserve Atlantic tunas among generally friendly states has produced poor results, for instance, leading states (including the two superpowers) were able to join forces to launch a successful regime for Antarctica during the height of the Cold War.

A sizable proportion of the success of environmental regimes is attributable to activities that are not regulatory in the ordinary sense. There is a strong tendency to think of regimes in regulatory terms. The promulgation and implementation of prescriptive regulations setting forth prohibitions, requirements, and permissions are important features of many regimes (Chayes and Chayes 1995). But these institutional arrangements regularly perform other functions as well (Young 1999a). Regimes may perform procedural functions (e.g., setting total allowable catches in fisheries on an annual basis or establishing and adjusting phaseout schedules for ozone depleting substances), and they often oversee programmatic activities (e.g., carrying out remedial action plans aimed at alleviating the effects of pollution in lakes or marine systems). Often overlooked is the function of regimes in identifying emerging problems, generating knowledge about the problems to be solved, and contributing to a shared understanding of the issues at stake among participating actors (Breitmeier, Young, and Zürn 2006).

Environmental regimes are dynamic in the sense that they change continually after their initial formation. Once established, institutional arrangements do not remain static over time. Environmental regimes wax and wane in terms of their capacity to solve problems. Some take on roles or are brought to bear in efforts to address problems that were not on the agenda at the time of their creation. It is possible to identify a number of patterns that constitute common pathways of institutional development and that generally take the form of emergent properties rather than planned developments (Young 2010). Some regimes (e.g., the ozone regime) go from strength to strength. Others (e.g., the Antarctic Treaty System) develop by fits and starts in a pattern of punctuated equilibrium. Still others (e.g., the climate regime) run into roadblocks that give rise to a pattern of arrested development.

The success of environmental regimes is highly sensitive to contextual factors. Context is a major determinant of the effectiveness of regimes. An arrangement that works perfectly well in one setting may fall flat in another setting. It is always important to think about scope conditions in assessing propositions about the effectiveness of environmental regimes. Some of the most notable features of the ozone regime, for instance, are unworkable in addressing the problem of climate change. This explains the importance of the propositions that we must go "beyond panaceas" in devising regimes to address real-world problems (Ostrom et al. 2007), and that it is essential to adopt a diagnostic approach in efforts to design regimes crafted to solve specific problems (Young 2002a; Ostrom et al. 2007; Young 2008).

Specific Findings

Beyond these general findings, regime analysis has generated a variety of more specific propositions about the effectiveness of regimes dealing with human-environment relations. Some of these findings are negative in the sense that they disconfirm popular notions about requirements for success. Others are positive, pointing to factors that are commonly associated with success.

Active participation on the part of a single dominant actor (commonly known in regime analysis as a hegemon) is not a necessary condition for success in solving international environmental problems. Dominant actors are important especially when they value a regime's products more than the cost of supplying them, making the relevant social system what is known as a privileged group (Olson 1965). But the absence of an engaged hegemon does not spell failure in this realm. What does seem important is the existence of a coalition of influential actors prepared to take the lead in jump-starting a regime at the outset and to provide an extra push at critical junctures along the road to success (Schelling 1978).

Success in the implementation of international regimes is likely to thrive on the establishment and maintenance of maximum winning coalitions rather than minimum winning coalitions. Regimes themselves are public goods, though individual members of the relevant groups of subjects may value them differently. In the extreme, some may regard them as public "bads." The fact that regimes are non-rival makes it desirable to maximize the size of coalitions supporting them rather than to form minimum winning coalitions of the sort common in domestic legislative settings (Riker 1962). Given the existence of the temptation to free ride, leading actors will have an incentive to make participation attractive to others rather than to minimize the number of those entitled to a share of the joint gains arising from the existence of an effective governance system. This is particularly so where regimes require ongoing implementation on the part of individual members.

The maintenance of feelings of fairness and legitimacy is important to effectiveness, especially in cases where success requires active participation on the part of the members of a sizable group over time. Neorealist perspectives suggest that both the formation and the implementation of regimes are about power—perhaps including soft power as well as hard power—all the way down (Strange 1982; Mearsheimer 1994/1995). The role of power is not only important in such settings, it is also a topic requiring more intensive analysis on the part of those interested in international regimes. But this does not eliminate the role of considerations of fairness and legitimacy (Franck 1990; Mattoo and Subramanian 2012; Coicaud and Warner 2013). Given the underdeveloped character of enforcement mechanisms at the international level, it is hard to elicit compliance on an ongoing basis from actors that do not accept a regime's prohibitions and requirements as broadly fair and legitimate.

Casting arrangements in the form of legally binding conventions or treaties does not ensure higher levels of compliance on the part of subjects. Many analysts assume that the "normative pull" associated with legally binding arrangements will have a positive effect on compliance. But the available evidence does not support this proposition (Breitmeier, Young, and Zürn 2006). Although hard-law arrangements may be desirable for other reasons, there is often a price to be paid for pursuing such arrangements not only in terms of the willingness of key actors to join regimes but also in terms of the depth of the substantive provisions they are willing to accept.

Arrangements featuring private governance and hybrid systems encompassing both public and private elements can solve some types of environmental problems. It is easy to exaggerate growth in the role of nonstate actors (e.g., multinational corporations, environmental NGOs) at the international level as well as the emergence of global civil society (Kaldor 2003). Nevertheless, while nation states remain core actors in international society, other actors are gaining ground. This opens up opportunities to solve problems through the development of hybrid systems (e.g., the system for classifying and insuring commercial ships) and even private regimes (e.g., the certification regimes for sustainable wood products and fish) rather than limiting the roles of nonstate actors to efforts to influence the operations of intergovernmental regimes (Delmas and Young 2009).

Multiple pathways can lead to success in efforts to solve many problems involving human-environment relations. It is generally a mistake to assume that there is one true path that must be identified and followed in efforts to solve specific environmental problems. Alternative solutions may vary in terms of other considerations, such as fairness or various notions of "good" governance. But what systems analysts call equifinality is a common phenomenon in the realm of environmental governance. This proposition applies with particular force to the selection of policy instruments (e.g., incentive systems vs. command-and-control regulations).

Findings about Institutional Interplay

Environmental regimes often interact both with one another and with regimes operating in other areas like trade and finance. The growth of interest in what has become known as institutional interplay is a recent development fueled by the observation that the number of distinct regimes operative in international society has grown rapidly in recent decades (Brown Weiss 1993; Underdal and Young 2004). A simple point of departure in thinking about interplay, pioneered by the long-term project on the Institutional Dimensions of Global Environmental Change, features two primary distinctions: one between horizontal and vertical interactions and the other between functional (or inadvertent) and political interactions (Young et al. 1999/2005). Because much of the responsibility for implementing the provisions of international regimes falls to their individual members, it is essential in thinking about effectiveness to consider vertical interplay (commonly referred to in recent literature as multi-level governance) as well as interactions among distinct institutional arrangements operative at the international level (Enderlein, Wälti, and Zürn 2010). Similarly, there is an important distinction between interplay that is largely unintended and often unforeseen and interplay involving intentional moves on the part of actors desiring either to manage interplay in order to promote problem-solving or to exploit interplay in order to advance their individual interests (Young 2002a).

Others have moved on from this point of departure. Particularly important in this regard are Stokke's account of the mechanisms of cognition, obligation, and utility maximization as determinants of the effects of interplay on problem-solving (Stokke 2001; Stokke 2011; Stokke 2012), and Raustiala and Victor's concept of institutional complexes as loosely coupled sets of non-hierarchically related arrangements operating in a single issue area (Raustiala and Victor 2004; Orsini, Morin, and Young 2013). Although the study of institutional interplay is a central concern at the domestic level, it constitutes a relatively new area of research at the international level. Nevertheless, some findings are already emerging from analyses of such matters (Oberthür and Gehring 2006; Oberthür and Stokke 2011).

Institutional interplay is just as likely to produce positive or even synergistic results as it is to lead to interference between or among regimes. This stream of analysis arose from a concern that tensions or even open conflict between or among distinct regimes would become an increasingly prominent feature of the institutional landscape in international society (Underdal and Young 2004). The logic underlying this concern is simple. As the number and variety of regimes operating in a given social space grow, the overlaps between and among them will increase. Since these overlaps are often unintended and commonly unforeseen in nature, it seems reasonable to expect that tensions will ensue (Lovejoy and Hannah 2005). But the research done so far on institutional interplay fails to confirm this expectation. Interactions may generate tensions. But institutional interplay often produces positive results and may even prove synergistic, as in the case of the regulation of ozone-depleting substances under the ozone regime that are also greenhouse gases and therefore sources of climate change (Oberthür and Gehring 2006; Velders et al. 2007; Oberthür 2009).

There is generally scope for resolving actual or potential conflicts between regimes through negotiations leading to mutual accommodation rather than by subordinating one regime to the other. For the most part, resolving such conflicts is not a matter of applying legal doctrines involving criteria like specificity and temporal sequencing to determine which regime should take precedence in the event of conflict between distinct arrangements. Rather, it is a matter of negotiating workable compromises that allow the regimes in question to operate effectively without undue interference in each other's domains (Oberthür and Stokke 2011). The most striking examples involve interplay between the global trade regime and a variety of multilateral environmental agreements involving the use of trade restrictions as a policy instrument (e.g., the regimes dealing with endangered species, hazardous wastes, the protection of the stratospheric ozone layer, and climate change) (Jinnah 2010; Faude and Gehring n.d.). The central challenge is to work out a modus vivendi allowing individual regimes to make progress toward solving the problems motivating their creation.

Regime complexes offer a way forward in situations that do not lend themselves to the creation of a single integrated governance system. Many issue areas (e.g., climate, biodiversity, marine pollution) feature networks of distinct regimes or "loosely coupled set[s] of specific regimes" that grow up in the absence of an overall blueprint (Keohane and Victor 2011: 7). Such complexes may range along a continuum from comprehensive and integrated governance systems for entire issue areas to total fragmentation (Keohane and Victor 2011). Regime complexes do not constitute a panacea. But they offer the advantage of being more flexible across issues and adaptable over time than more tightly coupled governance systems. Most observers have treated regime complexes as arrangements that arise spontaneously and evolve in the absence of deliberate interventions. But in some cases, these arrangements emerge at least in part as products of intentional initiatives that are appealing because they are easier to create than fully integrated systems, and because they are more resilient in the face of stresses occurring at the international or global level today. Regime complexes are likely to be common in many areas during the foreseeable future.

What Are the Cutting-Edge Issues Regarding Regime Effectiveness?

Taken together, these findings derived from hundreds of individual studies have significant implications for policy. It is worth bearing in mind, for instance, that not all regimes are regulatory in character, that legally binding arrangements are not always preferable to softer or more informal arrangements, that institutional interaction sometimes produces synergistic results, and that regime complexes may prove more successful than fully integrated regimes. Above all, it is critical to understand the problem of fit and, as a result, to discard hopes for panaceas and to sharpen the skills needed to engage in institutional diagnostics (Young 2013b).

At the same time, there is much more we can learn about the effectiveness of regimes dealing with human-environment relations that will be of interest to policy-makers as well as analysts. This section identifies a series of topics that constitute cutting-edge concerns in this field (Young, King, and Schroeder 2008; Biermann et al. 2009). There is no need to forge a consensus regarding the precise content of the research agenda. But it is helpful to get a sense of where we are headed regarding research on the effectiveness of environmental regimes.

Deep Structure

Regimes dealing with human-environment relations are specialized arrangements embedded in and reflecting the deep structure of international society (Underdal 2008). To be effective such arrangements must be broadly compatible with the essential features of the prevailing deep structure (Conca 2006). There is little point, for instance, in creating a regime complex for climate that requires the imposition of fundamental restrictions on the sovereignty of member states, or creating enforcement mechanisms that rely on severe sanctions to elicit compliance with the regime's rules. Yet it is easy to carry this line of thinking too far. The deep structure of international society is not static (Buzan 2004). The familiar power structure of the postwar era is shifting dramatically today. We need to recognize the growing importance of nonstate actors and the emergence of global civil society, and to think about the implications of these developments for perspectives built on the assumption that environmental governance is largely a matter of intergovernmental relations (Wapner 1997; Betsill and Corell 2007; Biermann and Pattberg 2012). So long as the normative gap is not too great, the development of innovative regimes can play a role in driving the evolution of the deep structure of international society. What we need to know in this realm is more about the constraints and opportunities associated with deep structure as they pertain to the operation of governance systems for specific problems like climate change and the protection of biodiversity.

Problem Structure

It is intuitively appealing to adopt the view that some environmental problems are harder to solve than others or, to use the terminology of the Oslo/Seattle team, that we can locate specific problems on a benign-malign spectrum (Miles et al. 2002). Climate change is certainly a more challenging problem than the depletion of the stratospheric ozone layer. But what exactly are the factors that make environmental problems harder or easier to solve, and can we devise a metric for assessing the nature of problems in these terms (Young 1999a)? Arild Underdal argues that environmental problems are hard to solve to the extent that they (i) are long-term policy problems with time lags between policy measures and effects, (ii) are embedded in highly complex systems clouded by uncertainties, and (iii) involve global collective goods not subject to single best effort solutions (Underdal 2010). These factors do pose important challenges for those seeking to solve environmental problems. They go some way, for example, toward explaining why it is so hard to come to grips with the problem of climate change. Yet there is considerable evidence to suggest that solutions to seemingly easy or benign problems can prove elusive, and that groups sometimes succeed in banding together to make serious efforts to tackle seemingly hard or malign problems. What we need to know in this realm is whether we are doomed to suffer the consequences of hard problems like climate change, or we can come up with innovative strategies to address such problems, given the emergence of effective leadership and the will to collaborate on the part of key actors.

Power

The role of power as a determinant of regime effectiveness is complex and contested, especially if we construe power to encompass soft power as well as hard power, cognitive power as well as structural power, and issue-specific power as well as general power (Nye 2011). Realist critics of regime analysis have often dismissed institutions as epiphenomena that reflect underlying distributions of power and that change as these distributions shift (Strange 1983). Those studying regimes sometimes seem to ignore or at least to downplay the role of power as a determinant of the capacity of these arrangements to solve problems. How can we come to terms with these diverging perspectives on the role of power? Regimes are embedded in overarching political orders, and they reflect the general presuppositions of political discourses dominant at the time of their creation. But this does not mean that they are of no significance in their own right, especially when treated as intervening forces that form links between the underlying drivers of human behavior and the outcomes flowing from human-environment interactions (Krasner 1983; Young 2002b). What we need to know in this connection is how to think about the role of power as a driving force in world affairs that does not blind us to the significance of other forces.

Breadth vs. Depth

Because participation in international environmental regimes is voluntary, there is a tendency to settle for arrangements that are shallow in terms of substance in order to make them palatable to all relevant actors. This is what Underdal and others have described as the law of the least ambitious program (Underdal 2002). The logic of those who advocate going forward even when commitments are shallow is that it is important to get the ball rolling and that institutional evolution will lead to a strengthening or deepening of commitments over time. This is an intuitively appealing argument, and examples like the regime for the protection of stratospheric ozone and (to a lesser extent) the regime dealing with long-range transboundary air pollution in Europe suggest that such a dynamic does occur under some conditions. But there is no reason to assume that such evolutionary processes will occur in all cases (Barrett 2003). The contrast between the regime for stratospheric ozone and the climate regime is striking in these terms. Not only has the stratospheric ozone regime proved effective in reducing drastically the production and consumption of ozone-depleting substances, it has also proven more effective in reducing emissions of greenhouse gases than the climate regime itself (Velders et al. 2007). What we need to know in this case centers on scope conditions. Under what conditions is it realistic to expect institutional evolution to work as a mechanism for deepening commitments in a manner required to achieve success in efforts to solve problems?

Compliance

We already know a lot about the sources of compliance (Young 1979; Mitchell 1994; Chayes and Chayes 1995; Raustiala and Slaughter 2002; Dai 2007). Because many observers regard compliance as the Achilles heel of international governance, however, this subject remains on the list of research priorities. The absence of a government in the ordinary sense in international society makes it hard to use sanctions—graduated or otherwise—effectively as a means of persuading or compelling those subject to a regime's rules to comply with their obligations. But this is not a fatal flaw (Young 1999a). In single best effort situations where one or a few actors can solve the problem once and for all, compliance is not a critical issue (Barrett 2007). Compliance is not a concern with regard to regimes that are not fundamentally regulatory in character. Even in regulatory settings, a management approach is sometimes more effective than an enforcement approach as a means of maximizing compliant behavior on the part of a regime's subjects (Chayes and Chayes 1995). Other factors, such as the extent to which subjects have engaged actively in the process of regime creation and the extent to which they feel that a regime constitutes a fair deal, can make a big difference in inducing actors to comply with a regime's rules and regulatory measures. What we need to know here is more about the sources of compliance (Victor, Raustiala, and Skolnikoff 1998). Because the emphasis at the international level must be on governance without government for the foreseeable future, we have a particular need to deepen our understanding of mechanisms that can produce compliant behavior in the absence of the sorts of sanctions we generally associate with the idea of enforcement (Skjaerseth, Stokke, and Wettestad 2006).

Fairness and Legitimacy

Despite the finding reported in the preceding section, there is substantial variation in the views analysts have expressed regarding the roles of fairness and legitimacy as determinants of the effectiveness of international regimes. Those who follow the logic of consequences and frame issues in collective-action terms have a tendency to dismiss or downplay the role of fairness, equity, legitimacy, and other normative factors in thinking about the success or failure of environmental regimes (Victor 2011). Those who think in terms of the logic of appropriateness and approach issues in social-practice terms, by contrast, are more receptive to the idea that such considerations are important determinants of effectiveness (Franck 1990; Franck 1995; March and Olsen 1998; Young 2001a). This divergence is not peculiar to the analysis of governance systems or regimes. It mirrors a larger and ongoing debate about the role of normative considerations as driving forces in international or global society. We do not need to adopt one or the other of these views in analyzing the effectiveness of regimes dealing with human-environment relations. What we do need, however, is an understanding of the conditions under which fairness and legitimacy are significant forces in this realm. This knowledge will have important implications for those designing regimes to address specific environmental problems, such as climate change or the loss of biological diversity.

Policy Instruments

The ideas of those who espouse incentive mechanisms, such as tradable catch shares or carbon taxes, in contrast to the more traditional mechanisms we generally lump under the heading of command-and-control regulations have dominated the discussion of policy instruments for several decades (Cole 2002). Clearly, the emphasis on such mechanisms has been salutary. Incentive mechanisms can alleviate the dynamic giving rise to the tragedy of the commons; they also can give subjects reason to focus on innovation on an ongoing basis. Yet it would be unfortunate if this were to lead to a situation in which one set of tools dominates our thinking about governance to the exclusion of others. There are important cases (e.g., climate change) in which it is difficult to make calculations regarding both the costs of leaving the problem unattended and the costs of taking effective action to alleviate the problem. There are also cases in which we have good reasons to override the use of discount rates of the sort commonly considered in conjunction with incentive mechanisms. It may make good sense in such cases to use command-and-control measures in place of or as a supplement to incentive mechanisms. Maintaining a well-stocked toolkit is clearly a good idea (Zaelke et al. 2005). What we need to know in this regard is more about the conditions under which specific policy instruments are likely to prove effective and how to make use of diagnostic procedures to bring this knowledge to bear on specific cases (Young 2008).

Interplay Management
Institutional interplay is on the rise. Whatever the attractions of creating comprehensive and integrated governance systems to address problems like climate change and the loss of biological diversity, we must prepare for a world that features rising levels of interplay between and among distinct regimes. As Keohane and Victor observe in their analysis of efforts to address climate change, relying on regime complexes or loosely coupled sets of specific regimes or regime elements dealing with broad and complex issues like climate change may prove advantageous in terms of flexibility across issues and adaptability over time (Keohane and Victor 2011). The implication of this proposition is that we must shift our attention from intensive studies of individual regimes to more expansive accounts of institutional interactions and especially regime complexes. Long familiar in domestic systems, this is a perspective that is relatively new at the level of international society. What we need to know here is more about the conditions leading to synergy rather than interference in institutional interactions and the conditions under which regime complexes produce flexibility and adaptability rather than chaos and confusion (Oberthür and Stokke 2011).

Nonlinear Processes

As we move deeper into a world of human-dominated ecosystems and the new planetary era now widely characterized as the Anthropocene, the need to improve our understanding of thresholds (often labeled tipping points) together with trigger mechanisms that can precipitate nonlinear changes has become urgent (Lenton et al. 2008; Carpenter et al. 2009). Nonlinear changes in socioecological systems are typically irreversible, sometimes abrupt, and often nasty from the perspective of human welfare. This makes it important not only to devise procedures to provide early warning regarding the onset of such changes but also to create governance systems able to adjust nimbly or agilely to the impacts of these changes. The trick is to create governance systems that are durable in the sense that they have the staying power to be effective combined with the adaptability to adjust quickly to changing circumstances. A regime that changes too readily in the face of modest stresses will not be effective. But a regime that is too rigid in the sense that it is unresponsive to major changes in the socioecological setting in which it operates will be vulnerable to forces leading to institutional collapse in a world in which nonlinear changes are common. What we need is a major step forward in our understanding of how to structure governance systems to maximize resilience while, at the same time, including procedures allowing for timely adjustments of the sort needed to maintain a good fit between socioecological conditions and institutional arrangements (Boyd and Folke 2011; Galaz et al. 2011).

Scale

The issue of scale in this context is a matter of the generalizability of findings regarding the effectiveness of governance systems across levels of social organization (Young et al. 1999/2005; Gibson, Ostrom, and Ahn 2000). To what extent do findings about issues like avoiding the tragedy of the commons derived from analyses of small-scale or local cases apply to comparable issues at the international level and vice versa (Keohane and Ostrom 1995; Dietz, Ostrom, and Stern 2003; Cash et al. 2006)? There are clear parallels between small-scale systems and large-scale or global systems with regard to the need to achieve governance without government (Young 2005). But there are also differences between these settings. Although many analysts use the term "international community" in discussing global issues, for example, there are major differences between local and global systems with regard to what is meant by the idea of community. What we need to know here is more about the limits of generalizability across levels of social organization regarding factors that determine the effectiveness of governance systems.

Moving toward Pasteur's Quadrant

Research on the determinants of effectiveness in regimes dealing with human-environment relations, much less the pursuit of sustainability in the Anthropocene, constitutes a young field. But it has already generated results of interest both to practitioners charged with administering regimes dealing with specific problems and to analysts seeking to understand the nature of governance, particularly in social settings where there is no government in the ordinary sense of the term. My own experience has convinced me that many rewards flow from a strategy of working back and forth between the worlds of analysis and praxis. It would be naïve to suppose that this line of research can reveal simple solutions to the great issues of our times, like controlling climate change and preventing the loss of biological diversity. But it would be equally inappropriate to dismiss the role of environmental regimes because they do not provide us with simple solutions to such overarching concerns. The way forward in efforts to enhance our understanding of the determinants of effectiveness is to make use of a suite of complementary modes of analysis. When the results converge, our confidence in the relevant findings rises. When they diverge, we are presented with puzzles of the sort on which science thrives. With persistence and a certain amount of good fortune, we will succeed in producing results that are of interest to analysts and practitioners alike and that, as a result, land us squarely in the domain of Pasteur's Quadrant (Stokes 1997; Clark 2007; Matson, Clark, and Andersson 2016).

II

Challenges of the Anthropocene

3

On Thresholds and Trigger Mechanisms

Introduction

A notable feature of complex systems is that they are subject to changes that are nonlinear, often abrupt, sometimes transformative, and commonly hard to anticipate. Interest in such changes occurring in biophysical systems and in the thresholds (or tipping points) and the trigger mechanisms associated with them has grown rapidly among natural scientists in recent years (Schellnhuber 2009). Ecologists analyze more or less abrupt state changes in which ecosystems flip from one state (e.g., a clear-water healthy lake) to another (e.g., a turbid-water lake characterized by hypoxic events and eutrophication) (Carpenter and Cottingham 2002; Carpenter 2003; Scheffer 2009). Climatologists seek to identify tipping elements in the Earth's climate system; they direct attention to situations in which seemingly modest trigger mechanisms could push this system past a point of no return leading to far-reaching and possibly abrupt changes in such forms as sea level rise, ocean acidification, or a dramatic increase in extreme weather events (Lenton et al. 2008). Earth system scientists have initiated the study of planetary boundaries in the sense of thresholds that must not be passed if the Earth system is to remain within a "safe operating space for humanity" (Rockström et al. 2009; Rockström and Klum 2015; Steffen et al. 2015). Both the locus of the points of no return and the nature of the trigger mechanisms that push the relevant systems past these thresholds are subject to considerable debate in specific cases. But no one doubts that such tipping points exist in a variety of biophysical systems or that passing them can lead to nonlinear and often transformative changes that seem disproportionate relative to the nature or intensity of the trigger events.

In this chapter, I explore parallel phenomena in social systems and in the socioecological systems that are central to the pursuit of sustainability in the Anthropocene.[1] Examples of social thresholds and trigger mechanisms are not difficult to identify. Think of cases like the collapse of the *ancien regime* in France following the storming of the Bastille on July 14, 1789; the onset of the Great Depression following the crash of the stock market on October 28–29, 1929; or the entry of the United States into World War II following the Japanese attack on Pearl Harbor on December 7, 1941. But we have made little effort to think rigorously or systematically about the nature of these social tipping points and trigger mechanisms, much less to understand the conditions governing their occurrence. My goal in this chapter is to take some steps toward filling this gap. To ground the discussion empirically, I introduce a range of examples returning often to: (i) transitions back and forth between conditions of peace and war, (ii) political shifts featuring the collapse of existing political systems and the formation of new ones, and (iii) economic transformations including the onset of economic depressions and of periods of sustained economic growth.

In the body of the chapter, I proceed as follows. The first substantive section addresses conceptual matters; it examines various aspects of the relevant universe of cases. The next two sections turn to the causes of social tipping points and trigger mechanisms and responses to them. This sets the stage for an analysis in the penultimate section that combines this account of social tipping points with parallel insights regarding biophysical tipping points to enhance our understanding of the dynamics of complex socioecological systems. The final section turns to the role of governance in this realm. Under what conditions can governance systems anticipate and even regulate the occurrence of thresholds? What roles might these institutional arrangements play in dealing with the consequences once a socioecological system passes a social or biophysical tipping point, minimizing harmful impacts, promoting positive outcomes, and stabilizing the resultant situation? An understanding of social thresholds and trigger mechanisms, I argue, will allow social scientists to make a more substantial contribution to the analysis of human-environment interactions and, in the process, help to illuminate the sorts of issues that have become a focus of interest among those interested in understanding the pursuit of sustainability in the new era we have come to know as the Anthropocene (Steffen et al. 2011; *The Economist* 2011).

Social Thresholds or Tipping Points

Most discussions of thresholds or tipping points are rooted in some form of systems analysis (Gunderson and Holling 2002; Steffen et al. 2004; Meadows 2008; Cornell et al. 2012; Wassman and Lenton 2012). A tipping point is a state of a system in which a relatively small event or trigger can precipitate nonlinear and typically disproportionate change in the system itself. Ordinarily, we think of these changes as irreversible. Once a system passes a point of no return, there is no going back to the status quo ante. A system may move toward some new equilibrium that is more or less stable. But it will not revert to its original state.[2] Some observers associate passing tipping points with transformative changes that are negative or disruptive. Common examples include the eutrophication of water bodies, the disruption of ecosystems caused by invasive species, and the collapse of glaciers and ice sheets. But negativity is not a necessary feature of tipping points. Passing a tipping point may result in a nonlinear change that most observers regard as positive as in the cases of a desert that experiences transformative change leading to the emergence of a savannah ecosystem, or an economy that achieves takeoff initiating a phase of sustained growth. More generally, assessing changes occurring in systems as positive or negative is a matter involving value judgments. The case of sustained economic growth provides a good example. Many regard economic growth as a good thing. From the point of view of climate change, however, growth looms as a source of the problem. Regardless of our evaluation of the outcomes, however, both those desiring to promote transformation and those seeking to prevent it are interested in systemic conditions that give rise to tipping points, together with the trigger mechanisms that can precipitate transformative changes once the systems of interest reach such thresholds.

Although most analyses of tipping points concentrate on the dynamics of biophysical systems, it makes sense intuitively to expect thresholds and trigger mechanisms to occur in social systems as well (Gladwell 2002). Economic bubbles burst. Companies go bankrupt. Political systems collapse. Wars break out. New social norms (e.g., the ban on smoking in public places, the acceptance of same-sex marriage) reach a point where they are able to drive out their predecessors. But can we identify such phenomena with sufficient precision to produce a tractable universe of cases that can support a research program aimed at framing testable hypotheses and developing empirical generalizations? In thinking about this question, I have found it helpful to differentiate three types of social thresholds and trigger mechanisms: explosions, cascades, and inflections.

An explosion occurs when a trigger event produces a flip of the system from one state to another fundamentally different state in a single abrupt step. The trigger is, in effect, a spark that ignites an explosion when it comes into contact with highly flammable material. Transitions from peace to war and back to peace often exemplify this type of situation. So do shifts in the character of political systems arising from critical elections and corporate collapses associated with bankruptcies. The assassination of the Austrian Archduke Franz Ferdinand in Sarajevo on June 28, 1914, precipitated a crisis leading within a few weeks to the eruption of violence between opposing sides that had been engaged in an arms race for some years. The result was a world war that drew a large number of states into violent conflict lasting for years, caused tens of millions of deaths, and led to the collapse of longstanding empires (e.g., the Austro-Hungarian Empire, the Ottoman Empire). The assassination of the archduke would not have led to the outbreak of war if the system had not been poised at a flash point at the time. But given the prevailing conditions, the assassination was enough to flip the system from peace to war within the span of a few weeks. The end of this war was equally abrupt. The fighting stopped precisely on November 11, 1918, when the armistice agreed to by the principal protagonists took effect.[3]

Similar flips can occur in political and economic systems. The American presidential election of 1932 replaced a conservative Republican administration with a progressive Democratic administration. The legislation the new administration introduced to deal with the impacts of the Great Depression during its first 100 days gave birth to what we now know as the New Deal and has few parallels in the annals of democratic political systems. The bursting of the real estate bubble in the United States during September 2008, to take another example, caused the sudden collapse of large financial services corporations (e.g., Lehman Brothers, Bear Stearns), which had been prominent features of the financial services landscape for decades and seemed to most observers like permanent fixtures. These corporations disappeared without a trace virtually overnight. Others escaped a similar fate solely through timely intervention on the part of the US Federal government under the terms of the Troubled Asset Relief Program (TARP) that George Bush signed into law on October 3, 2008.

A cascade differs from an explosion in the sense that it does not produce a state change overnight or in one decisive step. What distinguishes a cascade from an explosion is that the triggering event ignites a positive feedback process driving a system step-by-step from its initial state into some qualitatively different state over a period of time (Schelling 1978; Axelrod 1984). The idea of a domino effect applies to situations of this sort. Both economic depressions and epidemics are apt to exemplify this pattern. The crash of the US stock market on October 28–29, 1929, was a shocking event. But it did not produce a severe depression overnight or in one dramatic step. Rather, the initial crash triggered a chain of events in which market declines led to reductions in investments, producing a rise in unemployment, constraining purchasing power, and resulting in a vicious circle. Over a period of several years, this chain reaction spread from one country to another, reducing previously buoyant economies to straightened circumstances and causing intense human suffering in what became a worldwide depression by 1932–1933. These developments also enhanced popular support for the installation of authoritarian political regimes in several major countries, thereby playing a role in what became a cascade of events culminating in the outbreak of World War II.

The cascades giving rise to epidemics can be equally striking. As the proportion of a population affected by a contagious disease increases, the probability that the remaining members of the population will be affected rises. Under certain conditions, an outbreak that begins as a small relatively localized occurrence can trigger an accelerating process that ultimately affects large numbers of people worldwide. The Spanish flu pandemic that killed an estimated 40–60 million people during 1918–1919 remains one of the most dramatic instances of this sort of cascade (Kolata 2001). The fear associated with the recent outbreak of Ebola in West Africa arises from a concern that this virus could spread rapidly to other parts of the world. Despite the precautions that have been taken to prevent the occurrence of such runaway processes, public health officials warn that cascades of this type could occur on an even larger scale in the future (Galaz 2014). Similar phenomena occur in other realms, including more positive chain reactions like those underlying the growth of scientific knowledge and the remarkable expansion of corporations like Apple and Google. But in each case, the mechanism is the same. Non-linear change occurs in a series of (sometimes accelerating) steps driven by the mechanism of positive feedback. For those desiring to regulate such processes or simply to benefit from their occurrence, key questions concern the nature of the trigger mechanisms that set them off and the character of the feedback processes that drive them forward from one stage to another.

Inflections involve asymmetrical changes in key variables that alter the dynamics of complex systems in such a way that the systems experience transformative changes over time. Unlike the chain reactions involved in cascades, inflections occur when change in some critical element of a system is not matched by changes in other elements; they thus involve disconnects between key drivers rather than positive feedback processes. A prominent example centers on the shifting balance between mortality and fertility that produced the dramatic growth in human population occurring during the second half of the twentieth century (Bloom 2011). The mechanism underlying this transformative change is relatively simple. Advances in health care and the widespread use of pesticides (e.g., DDT) to control the spread of infectious diseases produced sharp reductions in mortality rates in many parts of the world. But fertility levels, driven by unrelated forces including engrained human attitudes, values, and social relationships, remained high for a considerable period of time. As a consequence of the resultant gap between fertility and mortality, many countries experienced net increases in their populations in the range of 3–4 percent annually. Fertility rates have since declined—slowly in some countries, more rapidly in others. Whereas the increase in the world's human population from 1 to 2 billion people took 120–125 years, the increase from 6 to 7 billion took just 12–13 years at the end of the twentieth century.

Somewhat similar inflections can arise from disconnects between demand and supply, leading to speculative bubbles that inflate rapidly until they reach a bursting point. Think of the tulip mania of the seventeenth century in this light. While supply was constrained by a number of factors including the availability of productive land and the nature of the annual growth cycle, there was nothing to limit the increase in demand.

Under such circumstances, runaway inflation in the price of a prized good can continue until some trigger event occurs that has the effect of pricking the bubble which then deflates more or less overnight.

Do explosions, cascades, and inflections constitute a tractable universe of cases that can support systematic investigation of thresholds and trigger mechanisms in complex social systems? I believe they do. But this is a question we cannot answer definitively on an a priori basis; sustained research will be needed to arrive at a convincing answer. In all three cases, nevertheless, it is possible to identify thresholds or tipping points in the sense of conditions that make systems susceptible to transformative change, along with trigger mechanisms that produce state changes or initiate processes leading to transformative changes over time. The fact that the three types of social thresholds and trigger mechanisms involve somewhat different dynamics may turn out to be a virtue rather than a limitation in this context. We want to understand the conditions under which a system experiences an explosive transformation flipping directly from one state to another (e.g., from peace to war) and to differentiate these conditions from those underlying the domino effect that constitutes the hallmark of cascades and the disconnects among key drivers that are the critical feature of inflections. But the important questions regarding each of these processes are similar. What conditions produce tipping points in the sense that they make a system fragile or brittle and create a setting in which processes eventuating in transformative change can get underway? What is the nature of the trigger mechanisms that ignite such processes? What can we do to prevent such changes when they seem undesirable or to promote them when they seem desirable? Taken together, we can think of these concerns as the *causal complexes* associated with social tipping points. Beyond this lies a constellation of questions relating to the impacts of transformative change and the institutional adjustments needed to maintain or even improve the quality of life for humans (individually and as members of social groups) in the wake of the changes that follow transformative processes. We can think of this second range of issues as encompassing *societal responses* associated with social thresholds and trigger mechanisms.

Causal Complexes

What can we say about the causes of the transformative changes that are the focus of this analysis of social tipping points? Two distinct sets of factors deserve consideration in this context. One set encompasses conditions that make a system susceptible to transformative change. I will analyze these conditions under the rubric of fragility. The more fragile the system, the less it takes to trigger transformative change (Schellnhuber and Held 2012). The other set involves the nature of the trigger mechanisms that initiate processes of change leading to explosions, cascades, and inflections. I will consider these forces under the rubric of intensity. The more intense the trigger, the more likely it is to generate transformative change.

Clearly, these factors interact with one another. When intense triggers impact fragile systems, it is to be expected that transformative changes will occur. Conversely, robust systems are able to tolerate mild triggers without experiencing any significant loss of resilience (Folke et al. 2004). The more interesting and more puzzling cases are those that are mixed in these terms. It takes very little to produce an explosion in a system that is extremely fragile. The assassination of the Austrian archduke in June 1914 may seem like a modest event relative to the magnitude and transformative character of the world war that ensued. But it is clear in retrospect that the military buildup in Europe and, in particular, the naval arms race between Germany and Great Britain had created an overarching situation that was so fragile that it required only a small spark to ignite a large-scale conflagration. Similar observations are in order regarding the outbreak of the American Civil War following the bombardment of Fort Sumter in Charleston harbor starting on April 12, 1861. On the other hand, resilient systems can withstand the impacts of intense trigger mechanisms without running a serious risk of experiencing transformative changes. While it is still hard to put all the events into perspective, the bursting of the real estate bubble, triggering the ensuing financial crisis in the United States and the worldwide recession of 2008–2009, seems like a relatively intense trigger. Yet the system proved resilient enough to withstand the impact of these events without experiencing transformative change. The crash of the stock market in October 1929 initiated a cascade leading to the Great Depression. But it is interesting to note that an even greater decline in the American stock market during the summer of 1921 failed to trigger a slide into depression.

Figure 3.1 depicts the relationship between fragility and intensity in more general terms, with the variable I call fragility on the horizontal axis and the variable I call intensity on the vertical axis. Setting aside important questions regarding measurement for the moment, I would place the French Revolution toward the northeast corner of the graph. By the late 1780s, the *ancien regime* was on its last legs, and the monarchy was unable to provide the leadership needed to sustain the system. The severe shocks associated with events culminating in the storming of the Bastille on July 14, 1789, were more than enough to topple the regime. The outbreak of World War I, by contrast, seems to me to lie farther toward the southeast corner. In this case, a relatively mild trigger was enough to produce an explosion bringing transformative change to an extremely fragile system. Compare this case with the impact of events like the October 1962 Cuban missile crisis, occurring during the height of the Cold War between the United States and its allies and the Soviet Union in the decades following the close of World War II. No matter how anxiety-producing this crisis was at the time, it appears in retrospect that there was enough resilience in the relationships underlying the Cold War to allow this system to survive relatively intense shocks without experiencing transformative

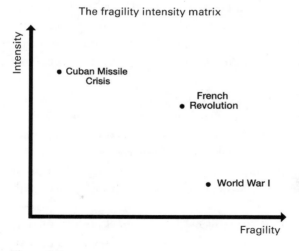

Figure 3.1

change. I would place this case farther toward the northwest in terms of the coordinates of figure 3.1.

Not surprisingly, this analysis presents a series of conceptual and operational challenges for those interested in developing the idea of social tipping points into a useful tool for analysis. First and perhaps foremost, is the matter of measurement. How can we determine the locus of social systems on a fragility scale ranging from highly resilient to extremely fragile? How can we assess the intensity of the shocks associated with various trigger mechanisms? Some crude distinctions along these lines do seem to make sense. Most observers would probably agree with assessments rating the *ancien regime* in France during the 1780s and the European politico-military system of the early years of the twentieth century as extremely fragile. But what about the regime in the Soviet Union in the years leading up to its collapse in 1991 and the regimes in places like Tunisia, Egypt, and Libya prior to the occurrence of the Arab Spring in 2011? It is probably fair to regard the basic economic system in place in the United States during 2008–2009 as quite resilient. But what made the American economy of 1929 more fragile than the same economy in 1921?

Similar remarks are in order regarding the intensity of trigger mechanisms. The shelling of Fort Sumter commencing on April 12, 1861, was a dramatic and symbolically significant event; it does not seem surprising that this event precipitated the outbreak of the Civil War. Much the same is true of the bombing of Pearl Harbor by the Japanese on December 7. 1941. By contrast, the events preceding the dissolution of the Soviet Union on December 26, 1991, do not seem exceptionally dramatic; most observers did not appreciate their significance at the time. If anything, the self-immolation of Mohammed Bouazizi in Tunisia on December 17, 2010, widely viewed as the trigger that ignited the Arab Spring, is even harder to interpret as an especially intense event. It is no wonder, under the circumstances, that authoritarian leaders in places like China worry about the possibility that some seemingly routine act of social protest could spark a cascade of events leading to a severe challenge to the stability of the regime.

There is much to be said for launching a sustained effort to devise criteria to be used in assessing the fragility of social systems and the intensity of trigger mechanisms. But we should not have inflated expectations regarding the results of this effort. In my judgment, the best we can hope for in this realm, at least during the foreseeable future, is the development of ordinal scales that allow us to say that some situations are more fragile than others and that some trigger events are more intense than others. It is possible that we can devise such scales encompassing 4 to 5 levels of fragility and intensity. This would be a relatively modest achievement. But even ordinal scales would be a big help in addressing the issues relating to social thresholds and trigger mechanisms under consideration here.

For one thing, such scales would allow us to think systematically about the fact that even close observers are routinely taken by surprise by transformative changes occurring in social systems. Many were concerned about the arms race that became a dominant feature of European international relations during the early years of the twentieth century. But few observers anticipated that the assassination of a ranking member of a royal family would produce the spark leading to a conflagration of the magnitude that erupted in August 1914 (Tuchman 1962). Almost no one—inside or outside the system—seems to have understood how fragile the Soviet system had become by the summer of 1991. Virtually everyone was unprepared for the sudden demise of the Soviet Union at the end of the year, followed by the emergence of the Russian Federation along with 15 additional independent countries in place of the Union of Soviet Socialist Republics. Perhaps even more surprising is the unfolding of what we now call the Arab Spring during the early months of 2011. And while economic bubbles seem remarkably fragile—even somewhat preposterous—in retrospect, it is extraordinary how blind observers are to the fragility of such situations (e.g., the tulip mania of the seventeenth century, the South Sea bubble of the eighteenth century, the dot.com bubble at the end of the twentieth century) right up to the moment of bursting. Even the sharpest observers failed to understand the artificiality of the real estate bubble in the United States during the early years of this century, a fact that seems extraordinarily difficult to explain in retrospect.

A distinctive feature of human affairs arises from deliberate efforts on the part of some individuals and groups to push social systems toward tipping points and to activate trigger mechanisms that precipitate transformative change, while others who are committed to the status quo work just as hard to prevent such initiatives from succeeding. This phenomenon is particularly prominent in cases involving revolutionary changes in political systems. The competition between revolutionaries and conservatives went on for years in eighteenth-century France prior to the storming of the Bastille on July 14, 1789, commonly regarded as the trigger event initiating the cascade eventuating in the collapse of the monarchy and with it the *ancien regime*. The struggle between revolutionaries and conservatives in Russia was protracted, including the revolutionary events of 1905 that played some role in preparing the way for the cascade of events leading to the collapse of the czarist regime during the fall of 1917. In other cases, intentional behavior is less clear-cut but still a significant factor in the occurrence of social explosions, cascades, and inflections. It is probably accurate to say that those responsible for the assassination of Franz Ferdinand in Sarajevo in June 1914 did not set out deliberately to ignite the conflagration of World War I. But some have argued that Franklin Roosevelt was on the lookout for an event that could galvanize American public opinion in support of the entry of the United States into World War II and that he viewed the Japanese attack on Pearl Harbor through this lens. There is little reason to suspect that those desiring to undermine the capitalist system played important roles in causing the economic crashes of October 1929 and October 2008. Yet it seems clear that anti-capitalist forces have worked hard to increase the fragility of some economic systems and, in the process, played a role in moving them toward what can be described as tipping points.

A remarkable feature of these processes is that those struggling to undermine prevailing economic, political, and social systems can work hard for long periods of time with little apparent success but take actions at critical junctures that produce seemingly disproportionate results. Opponents of the *ancien regime* struggled for years—even decades—with little apparent success; the same is true of revolutionaries in Europe during the nineteenth and early twentieth centuries. Many leaders of the international socialist and communist movements of this era regarded Russia as a much less suitable environment for revolutionary activities than more industrialized countries like Germany. To take a very different example, the contemporary campaigns in the United States to ban smoking in public places and to gain acceptance for same-sex marriage seemed to make relatively little headway for years before reaching critical thresholds where modest additional efforts produced dramatic results. At this stage, our understanding of such processes is limited. It may well be that epidemiological models can help to account for such phenomena. Something like the process ecologists have described in analyzing the spread of invasive species in ponds may be at work here. If the area covered by an invasive species doubles every day, it will take only one day to go from a situation in which the invasive species is present in only half the pond to a situation in which it covers the whole pond (Brown 1978). But there can be little doubt that the social phenomena under consideration here are highly complex; our efforts to understand them better and to reduce the frequency with which we are taken by surprise by sudden transformations remain at an early stage.

Societal Responses

Tipping points are points of no return. Once a social system crosses such a threshold, the ensuing changes become irreversible. There is no going back to the status quo ante. But there are striking differences in the nature of the results going forward. Some cases feature clean breaks with the past; the new order bears little or no resemblance to the prior order. Other cases feature dynamics that allow for the emergence of new orders that share a number of the defining features of the old orders. Still other cases produce more mixed results, leading to breaks with the past that are cleaner in some respects than in others. And, of course, there is no basis for assuming that new orders arising in the wake of transformative changes will prove stable or resilient over time. Some state changes yield outcomes that are highly stable; others give rise to sequences of events that include the onset of new tipping points and the sowing of the seeds of the next transformative changes.

A consideration of some concrete cases will help to flesh out these general observations. The collapse of the Soviet Union at the end of 1991 produced a decisive break in political terms. There is no prospect of resurrecting the Soviet state; the reunification of the republics that made up the Union of Soviet Socialist Republics is not an option, and the return of a fully socialized or state run economy is not likely any time soon. Still, it is interesting to note that the collapse of the Soviet Union did not put an end to authoritarian politics in Russia or to the prominence of economic actors (e.g., Gazprom, Rosneft) in which the state has a controlling interest. Some would argue that the emergence of a fully democratic political system in Russia would require deeper changes in the national culture. There can be no doubt about the transformative character of the changes brought about by the French Revolution. The monarchy collapsed, the class structure associated with the old order underwent profound change, and the rights of ordinary citizens emerged as important elements of the political culture. Yet it is also worth noting that the monarchy reemerged, albeit in a greatly weakened form, in the aftermath of the Napoleonic Era in France. The decisive shift to a republican form of government in France did not occur until the aftermath of the Franco-Prussian War in 1870–1871.

Similar observations are in order regarding the transformative changes associated with major wars. Both the Austro-Hungarian Empire and the Ottoman Empire collapsed in the wake of World War I, never to rise again. The Russian Revolution of 1917, which brought an end to the imperial order in Russia, was made possible in part by the impacts of World War I. Yet many have argued that the Treaty of Versailles (whose signing on June 28, 1919 marked the formal close of World War I) paved the way for the onset of World War II by leaving the basic features of the European political order intact and inspiring the losers to make a concerted effort to reverse the outcome of the war (Keynes 1920). As a result, some observers see the two world wars as elements in a set of linked developments in the twentieth century that ended only in the aftermath of World War II, with the creation of democratic political systems in Germany and Japan, the establishment of the European Union, and the initiation of a process of decolonization leading to the end of the British and French Empires. In some respects, the American Revolution is a cleaner case in terms of this discussion of irreversible change. Triggered by the battles of Lexington and Concord on April 19, 1775, the ensuing war set in motion a series of events that resulted in the establishment of the United States as an independent state and the emergence over time of the United States as a powerful force in international society. Still, it is worth noting in this context that the establishment of a federal government in the United States following the Revolution was by no means a foregone conclusion and that the war did not put an end to the British Empire (Maier 2011; Stewart 2015). In many ways, the growth of British imperial power reached a peak during the nineteenth century, giving rise to the idea of Pax Britannica.

The economic transformations I have referred to in earlier sections offer several additional insights regarding the consequences of crossing the thresholds associated with tipping points. The Great Depression of the 1930s, which took the form of a cascade triggered by the stock market crash of October 28–29, 1929, was almost certainly the most disruptive worldwide economic event of the twentieth century. Yet it did not put an end to capitalism or the capitalist economic order. The economic systems in place in the advanced industrial societies of today are not qualitatively different than those in place prior to the Great Depression.[4] For their part, rising powers like China and India have begun to embrace key features of capitalism as a form of economic order. What does seem to have undergone irreversible change as a result of the Great Depression is the allocation of social functions between the private sector and the public sector. Even the most conservative actors in contemporary societies are not calling for a fully-fledged reversal of the enhanced role of the state in the areas of health, education, and welfare. These comments also serve to put the "great recession" beginning in fall of 2008 into sharper perspective. While it would be wrong to underestimate the changes brought about by this recession, it is difficult to construct a compelling case for the proposition that this was a transformative event on a macro scale. Governments have made some efforts to regulate the practices (e.g., the creation of financial derivatives) widely seen as causes of the recession. But the effectiveness of these efforts is open to serious doubt. In the longer run, we may see this episode more as an indicator of the relative economic decline of the United States and Europe and the rise of other powerful economic players than as a transformative event with regard to the nature of the prevailing economic order.

What is certainly true about the transformative changes resulting from social tipping points and trigger mechanisms, however, is that they often produce winners and losers. The old aristocracy in Europe lost much of its influence in a cascade of developments starting with the French Revolution and the political reforms of the nineteenth century and continuing with the Russian Revolution and the rise of the Soviet state along with the decline of empires brought about by the two world wars. The United States and the Soviet Union emerged from the trials of World War II as global powers whose capacity to exert influence greatly exceeded that of Germany and Japan as the losers of the war,

Britain and France as exhausted victors, and China as a society wracked by civil strife. In somewhat more mundane terms, it is worth noting the effects of economic transformations on the fortunes of individuals. Truly wealthy individuals are often able to ride out recessions and even depressions with their economic positions intact. What the Great Depression of the 1930s and, to a lesser extent, the recent great recession did was to undermine the economic standing of many members of the middle class (Piketty 2014).

In any case, there can be no doubt about the impacts of transformative change on the economic, political, and social standing of various players in social systems. Some will see this in a positive light as a mechanism that drives the circulation of elites, thereby preventing societies from stagnating as dominant players become increasingly attached to the status quo and unwilling to accept the risks associated with the pursuit of economic and political innovations. Others will take a more negative view, arguing that the dramatic changes in the fortunes of various players associated with transformative systemic changes are illegitimate and apt to lead to a decline in the social trust or sense of community that is such an important source of order in most societies. But whatever our thinking in these terms, there is no denying that the new orders emerging in the wake of transformative economic, political, and social changes will become, in their turn, sites of competition between defenders of these new orders and a new generation of radicals bent on overthrowing or at least reforming these orders dramatically. This is the principal source of dynamism in social systems. The new orders may be fragile or robust. But even in the most robust cases, fragility is likely to rise as the generation that presided over the transformation leaves the scene and uninspired conservatism becomes a more powerful force. Triggering events may occur sooner or later, especially when we expand our scope to consider biophysical triggers of the sort discussed in the next section. But the overall conclusion is clear. The transformative changes that occur once a tipping point is passed do not land a system in a state of stable equilibrium that can be counted on to remain in place for the indefinite future. Although some outcomes are more stable than others, social movements calling for more or less drastic changes will arise sooner or later once a new order comes into existence.

Thresholds and Trigger Mechanisms in Socioecological Systems

Turn now to a consideration of the coupled systems that have become a focus of attention among those seeking to understand and come to terms with the onset of the Great Acceleration that has given rise to the era we often refer to as the Anthropocene (Steffen et al. 2004). These are systems in which anthropogenic forces play important roles in the dynamics of biophysical systems, while biophysical forces continue to act as major determinants of the development of social systems (Vitousek et al. 1997). There is nothing new about these relationships. Biophysical forces have shaped social systems from the outset (Ponting 2007). On a macro scale, in fact, it seems clear that the relatively benign character of the Earth's climate system over the last 10,000 years—the era known as the Holocene—has played a key role in providing a setting conducive to the development of settled agrarian and industrial societies. The increasing turbulence that climate change is expected to produce in the future is a source of understandable concern in these terms. Human actions have transformed biophysical settings, at least on a small scale, for millennia (Turner et al. 1990). Think of the deliberate use of fire to manage ecosystems as a case in point. What is new is the capacity of humans to alter biophysical systems on a planetary scale. Climate change and the loss of biological diversity are perhaps the most dramatic current cases in point (Kolbert 2014). But these cases are only illustrative of a fundamental change in the character of human-environment interactions on a planetary scale.

With regard to the issues addressed in this chapter, several sets of interactive processes are worth distinguishing and considering separately. Biophysical forces can contribute to the fragility of social systems; they can also serve as trigger mechanisms causing social systems to cross the thresholds associated with tipping points. Conversely, social forces can play a role in increasing the fragility of biophysical systems; they can also produce triggering events that set in motion processes leading to transformative changes in biophysical systems. Of course, these events can also coalesce to produce complex interactive processes. Biophysical triggers can initiate transformative social changes that then have dramatic consequences for biophysical systems and so forth in an interactive cycle that is difficult to grasp clearly, much less to control effectively.

Biophysical forces figure prominently as determinants of the fragility or robustness of social systems. This is particularly easy to see in the case of subsistence systems finely tuned to specific natural conditions. Small communities vulnerable to coastal erosion or heavily dependent on a specific stock of living resources are familiar examples. As anthropologists have often noted, some of these communities have found ways to achieve a measure of resilience by developing a capacity to adapt nimbly or agilely to changing biophysical circumstances. But this is not always the case, especially when communities become rooted in specific places and lose the capacity to pull up stakes and move to new locations when biophysical threats to their well-being become severe (Diamond 2005). It is easy to see that modern societies are by no means invulnerable to biophysical threats to their robustness. Consider the example of the coastal communities in Japan that had no way to defend themselves against the impacts of the March 11, 2011 Tohoku earthquake and tsunami.

As this last example suggests, extreme events like hurricanes, earthquakes, and volcanic eruptions can act as trigger mechanisms initiating transformative changes in social systems. The December 26, 2004, Sumatra-Andaman earthquake and tsunami killed over 200,000 people and devastated communities across South Asia. Hurricane Katrina, which made landfall in Louisiana on August 29, 2005, produced destruction from which the city of New Orleans has yet to recover fully. The March 2011 earthquake and tsunami not only knocked out the Fukushima Dai-ichi Nuclear Power Plant; it also triggered a far-reaching reexamination of energy policy in Japan. As these cases suggest, the transformative consequences of these triggering events are at least partly manmade. Many observers have pointed out that it was known well in advance that the dikes protecting New Orleans could not withstand the impact of a direct hit from a storm with the intensity of Katrina but that no one was willing or able to make the investments required to address this problem.[5] Similar observations are in order regarding the Fukushima disaster. It has become clear in the aftermath of the March 2011 Tohoku earthquake and tsunami that pre-existing conditions at the nuclear power plant left a great deal to be desired in terms of vulnerability to an extreme event of this magnitude. It turns out, therefore, that fragility in such cases is often at least partially a social phenomenon that does not become apparent until some extreme biophysical event draws attention to what should have been well understood in advance.

What has become apparent more recently is that human actions can play a key role in determining the fragility of biophysical systems, even acting as trigger mechanisms leading to transformative changes in some large-scale systems. Dramatic examples occurring in small-scale and medium-scale settings are not difficult to find. The actions of the human residents of Easter Island are thought to have played a major role in increasing the fragility of the island's ecosystem, with results that proved disastrous in social as well as biophysical terms (Diamond 2005). It seems clear that agricultural practices poorly suited to prevailing ecological conditions played an important role in increasing the vulnerability of biophysical systems to the conditions we now describe as the Dust Bowl of the 1930s in the American Southwest (Egan 2006). What is new in this realm is the capacity of human actions to increase the fragility of biophysical systems on a planetary scale. The most dramatic example today involves the Earth's climate system and the onset of what we now describe as "dangerous anthropogenic interference" in the operation of this system.[6] But, as the idea of the Anthropocene makes clear, we are living now in a world of human-dominated ecosystems in which the actions of human beings are often critical determinants of the fragility of large-scale biophysical systems.

What about anthropogenic triggers of transformative changes in biophysical systems? In some cases, such triggers seem relatively easy to identify. The invention of the Watt steam engine, which dates from the issuance in 1769 of a patent for a separate condenser connected to a cylinder by a valve, marked a turning point in what we think of in hindsight as the Industrial Revolution. Among other things, this development played a role in the emergence of England as the frontrunner in the transformation of economic processes in Europe and North America during the nineteenth century, and in the transformation of biophysical systems arising from the search for and use of fossil fuels to power industrial systems. The invention and deployment of the explosive harpoon and the harpoon cannon in the closing decades of the nineteenth century initiated a cascade of events leading to severe depletions of populations of great whales on a global scale. The start of electrical power production at the Obninsk Nuclear Power Plant in the USSR on June 27, 1954, ushered in the era of what we sometimes call peaceful uses of nuclear energy with consequences for biophysical systems that remain highly controversial today. Other anthropogenic triggers are harder to isolate but may produce consequences that turn out to be profound over time. For example, it is possible that "anthropogenic interference in the climate system" has already pushed the Greenland ice sheet past a point of no return, ensuring that it will disintegrate in the future. But this process may take several centuries, and we are not likely to be able to determine whether or when we pass(ed) a point of no return regarding the future of the Greenland ice sheet for a long time to come.

As this discussion makes clear, there are powerful interaction effects between biophysical systems and social systems with regard not only to the fragility of socioecological systems but also to trigger mechanisms that can push these systems beyond points of no return. Humans bring invasive species to new areas (e.g., rabbits in Australia, zebra mussels to the Great Lakes), which thrive in ways that have profound impacts on natural systems, producing in turn far-reaching consequences for the activities of humans. Conversely, the Little Ice Age in Europe, covering the period roughly from 1550 to 1850, had profound impacts on the viability of agrarian communities and on the political systems of the pre-modern era.[7] But it also played a role in the search for new fuel sources (e.g., coal) and the economic processes that eventuated in industrialization with its increasing impacts on biophysical systems on a larger and larger scale. Of course, the most dramatic contemporary example involves climate change, as it now seems clear that human actions are causing changes in the Earth's climate system with effects that are likely to bring about profound changes in social systems over time as well as further human actions, such as various forms of geoengineering, affecting the climate system (Keith 2013; National Research Council 2015a and 2015b). From the perspective of management or governance, it is the occurrence of these interaction effects that gives rise to increasingly complex systems and, as a result, the most far-reaching challenges of achieving sustainability in the current era. Can we take steps to avoid human actions that trigger large-scale disruptions in biophysical systems? Are there ways to minimize the social costs or even to take advantage of transformative changes in biophysical systems occurring on a very large, increasingly planetary, scale?

The Role of Governance

Governance, as I use the term, is a social function centered on efforts to steer societies toward socially desirable outcomes and away from socially undesirable outcomes (Ch. 1 *supra;* Young 1999; Young 2013a). Addressing collective-action problems, like the depletion or exhaustion of common pool resources, is a paradigmatic example of the effort to meet needs for governance. We are apt to think of material entities like governments as the agents responsible for addressing such matters. But it is important to note that the existence of a government in the ordinary sense of the term is neither sufficient to ensure that such needs will be met nor necessary to perform this role. Lack of capacity, institutional arthritis, widespread corruption, or the rise of authoritarian leaders may prevent governments from meeting even the most basic needs for governance in specific situations. Perhaps more interesting in the context of this discussion is the observation that situations featuring "governance without government" are not uncommon. Small-scale societies often develop social norms and cultural practices that are effective in avoiding traps like the tragedy of the commons in the use of common pool resources (e.g., stocks of fish, stands of trees) important to their ways of life (Ostrom et al. 2002; Dietz, Ostrom, and Stern 2003). There are even cases of success at the international level in efforts to solve problems that feature governance without government, as the case of phasing out the production and consumption of ozone-depleting substances on a global level makes clear (Benedick 1998). A key question confronting us today centers on the prospects for coming to terms with far-reaching problems like climate change in the absence of anything resembling a world government. In effect, we need to determine the conditions under which it is possible to meet needs for governance on a large scale without resorting to the establishment of at least a rudimentary form of world government (Dietz, Ostrom, and Stern 2003; Biermann 2007; Young 2013a).

Note also that the relevant challenge may involve efforts either to prevent a system from crossing a threshold precipitating transformative change or, conversely, to push a system past such a threshold in the interests of initiating transformative change. It makes sense to assume that those in positions of authority will work hard to maintain their own positions and therefore take any steps needed to avoid crossing thresholds leading to the collapse of existing political orders. Much the same is true regarding existing economic orders. There are few winners from economic disruptions of the magnitude exemplified by the Great Depression of the 1930s or even the great recession of 2008–2010, though even in such cases those desiring to bring down the capitalist order may be happy to see the occurrence of events like the stock market crash of October 1929. The situation regarding the outbreak of war is more complicated, since there are sometimes actors who believe that they will emerge as victors and who knowingly take steps likely to trigger the onset of war, though in reality their expectations often turn out to be based on little more than fantasies. Efforts to trigger the onset of sustained economic growth, on the other hand, constitute another story. Policy-makers and those in charge of administering economic assistance programs often focus on the idea that it is possible to cross a threshold beyond which an economic system achieves a capacity to propel itself from strength to strength in a manner that produces what proponents like to call sustained economic growth (Rostow 1991). Of course, many "theories" about the nature of the relevant tipping points and trigger mechanisms turn out to be erroneous, and we now know that economic growth is not an unmitigated blessing, especially when treated as an end in itself. Nevertheless, this is a realm in which actors frequently think of the role of governance as a matter of identifying tipping points and taking actions to push systems across key thresholds rather than as a matter of taking whatever steps are required to avoid crossing a threshold leading to transformative change.

However the objectives of governance are framed in specific cases, we can identify several distinct governance functions relating to tipping points, trigger mechanisms, and the situations arising in the aftermath of transformative changes (Young 2012b). First, and in some ways foremost, is the matter of monitoring, assessment, and early warning. A striking observation about social tipping points and the ensuing transformative changes is the frequency with which they take all parties concerned by surprise. This is a matter, on the one hand, of an inability to gauge the fragility of social systems and therefore to recognize how little it may take to trigger transformative changes and, on the other hand, of an inability to anticipate the occurrence of trigger events and to gauge their intensity. Not only do we lack clear indicators of fragility; there also appears to be unwillingness in many quarters to think rigorously about the fragility of existing orders. This is due apparently to a number of factors, including a lack of creative ideas about the development of indicators, a fear that focusing too much attention on such matters will be interpreted as a sign of disloyalty to those in charge of the existing order, a natural tendency to avoid thinking about what may seem like low probability events, and just plain wishful thinking. Still, it seems critical for those in charge of governance systems to devote more attention to understanding fragility and the signs of advancing fragility, if only as a matter of self-interest on the part of current elites.

The issue of anticipating the occurrence of trigger events is of a somewhat different nature. Part of the problem clearly lies in the inherent unpredictability of some trigger mechanisms. We lack the capacity to predict the timing and location of major earthquakes, for example, and it is by no means clear that more research will alleviate, much less eliminate, this problem. It is equally clear that we lack the capacity to predict the occurrence of political collapses (e.g., the collapse of the Soviet Union at the end of 1991) and the bursting of economic bubbles (e.g., the bursting of the real estate bubble in the United States in the fall of 2008). There is also a systematic bias toward underestimating the relevance of low probability events that may occur at some unspecified time in the future (Kahneman 2011). But none of this alters the proposition that allocating more time and attention to matters of monitoring, assessment, and early warning must be treated as a matter of the highest priority for those who are interested in managing tipping points and trigger mechanisms leading to transformative changes in social or socioecological systems.

We come next to the roles that governance systems can and do play in preventing or encouraging trigger events that push systems across thresholds leading to transformative changes. The options here vary depending upon the extent to which the relevant processes take the form of explosions, cascades, or inflections. In the case of explosions, which feature dramatic and sudden flips from one state to another, the most effective strategy will involve efforts to reduce or increase the fragility of the relevant system. Whereas the trigger mechanisms are hard to control, the likelihood that any given trigger event will produce an explosion is a function of the fragility or robustness of the relevant system. The cases of cascades and inflections differ in this regard. Quick and decisive actions following the start of a sequence that can initiate a cascade may play an important role in determining the course of events. The differences in the reactions of the US Federal government to the stock market crashes of October 1929 and October 2008 are illustrative. On the other hand, an exaggerated fear of the domino effect can lead to inappropriate responses to events that may seem to be pushing a system past a point of no return. One interpretation of the Gulf of Tonkin incident on August 2, 1964, for example, is that American policy-makers overreacted to a murky situation due, in part, to an exaggerated fear that the governments of the states of Southeast Asia would fall one after another like a string of dominoes once the process got underway. In the case of inflections, the trick is to identify the variables involved and to take steps to bring them into alignment as quickly as possible. With regard to population, for instance, avoiding dramatic growth in population would have required focused efforts to reduce fertility to match the decline in mortality associated with the use of pesticides and improved health care practices in the later part of the twentieth century. Whatever its faults, which are hotly debated, the one-child policy in China, introduced formally in an Open Letter from the Central Committee of the Communist Party of China dated September 26, 1980, was based on the premise that an imbalance of fertility and mortality would lead to a dramatic increase in China's population, which in turn would undermine efforts to lift a large fraction of China's people out of poverty (Greenhalgh 2008).

A third role for governance in this realm involves actions intended to stabilize or rebalance the situation following the occurrence of transformative change in the relevant system. In the case of war, this may mean working to restore peace on terms that all the participants find acceptable. In the wake of a revolution that brings down an old political order, on the other hand, the challenge is to create and stabilize the new order. In the aftermath of the American Revolution, for example, this effort led to the Constitutional Convention held during the summer of 1787 and the launching of the United States as a politically unified country with a national government possessing considerable authority, in contrast to a loose consortium of largely independent states (Stewart 2008; Stewart 2015). Following the collapse of the Soviet Union at the end of 1991, by contrast, a critical challenge for governance was to devise a means of dismantling the socialist economic system and, in the process, to transfer ownership of most of the means of production into private hands. As this case makes clear, there is ample room for the occurrence of governance failures in dealing with the aftermath of transformative change. Other examples feature less disruptive but no less profound transitions in the state of social systems. For instance, the Nineteenth Amendment to the US Constitution giving women the right to vote, which took effect on August 18, 1920, did not produce a revolution in the ordinary sense of the term. But it more or less doubled the number of adults eligible to vote in US federal elections at a single stroke and required a concerted effort on the part of the governance system to make the adjustments needed to accommodate this change. Given the fact that the distribution of political preferences among women often differs significantly from the distribution of preferences among men, it is easy to see that the consequences of this transformative change have been profound.

Concluding Observations

Tipping points and the trigger mechanisms that push systems across critical thresholds initiating transformative changes are no less important in social systems than they are in biophysical systems. When we turn to socioecological systems, which feature complex interactions between biophysical and anthropogenic forces, the challenges of dealing with thresholds, trigger mechanisms, and transformative changes are profound. A major problem in addressing these challenges arises from the facts that we have a limited ability to assess the fragility of the relevant systems and that we find it hard to anticipate both the occurrence and the intensity of trigger events. The problem here is by no means a matter of the deficiency of social science in contrast to natural science. It is just as hard to evaluate the fragility of the Earth's climate system, for example, as it is to determine the fragility of the global economic system in place today. There is little reason to expect that we can eliminate critical uncertainties regarding these matters during the foreseeable future. Decision-making under uncertainty will remain the order of the day, whether we are trying to avoid "dangerous anthropogenic interference in the climate system" or to prevent a great recession like the one occurring in 2008–2010 from degenerating into a global depression of the sort that occurred in the 1930s. Still, this does not mean that we are helpless in the face of these large-scale forces. There is every reason to invest in a concerted effort to monitor and assess these processes and to put in place rapid response capabilities to deal with the problems arising once we move beyond points of no return in socioeconomic, biophysical, and especially socio-ecological systems.

4

The Sustainability Transition

Introduction

There is a growing gap between reality and our guiding assumptions about the behavior of socioecological systems that threatens to undermine efforts to create and administer governance systems capable of steering human-environment interactions toward sustainable outcomes in the Anthropocene. The dominant paradigm regarding such matters assumes—implicitly if not explicitly—that the socioecological systems of interest are characterized by change that is linear, gradual, and foreseeable rather than discontinuous, abrupt, and surprising. We assume that populations of renewable resources (e.g., fish stocks) will grow or decline smoothly and relatively slowly, so that we do not have to worry about dramatic and sudden collapses once thresholds are crossed. Similarly, we assume that temperatures will increase and sea levels will rise gradually and uniformly as a result of climate change, so that we will be able to adapt to these changes on an incremental and generalized basis when we get around to taking climate change seriously. For the most part, we also assume that human responses to biophysical changes are linear and gradual. We tend to take it for granted, for instance, that economic adaptation to climate change will be incremental in nature and that policy processes relating to other environmental problems will not involve discontinuous and surprising institutional changes once some sort of crisis erupts. When dramatic or abrupt changes do occur, we are taken by surprise. Such surprises are particularly prevalent in cases where telecoupling produces changes in which drivers are far removed spatially or functionally from their effects.

From a purely analytic perspective, these assumptions have significant advantages. The tools we have developed to organize thinking about systems in which change is linear and gradual and cause/effect connections are tight are far more sophisticated than the tools available for understanding complex systems featuring thresholds, telecouplings, abrupt changes, and emergent properties that generate novel challenges for governance. The problem is that there is a large and increasingly important gap between the resultant models and reality in these terms. This is a consequence, above all, of the emergence of complex systems in which anthropogenic drivers are powerful forces on a large scale.

Human actions have produced significant impacts on biophysical systems for thousands of years (Turner et al. 1990). But the scope and magnitude of the human forces operating in socioecological systems up to and including the Earth system itself have risen dramatically (Steffen et al. 2004; Schellnhuber et al. 2004), leading prominent scientists to conclude that we have entered a world of human-dominated ecosystems (Vitousek et al. 1997) and even that the planet has transitioned from the Holocene into a new era best described as the Anthropocene in view of the profound impact of human actions on a planetary scale (Crutzen and Stoermer 2000, Crutzen 2002; Steffen et al. 2011). Dealing with biophysical systems or socioeconomic systems alone is challenging enough. But the resultant socioecological systems are far more complex than anything we have encountered previously; there is every reason to expect these conditions to intensify over time. It follows that we must learn to live with nonlinearities and the inevitable uncertainties associated with the prominence of emergent properties in complex systems, if we are to build effective institutions to govern human-environment interactions in the interests of achieving sustainability in the Anthropocene.

The main message of this chapter is that reliance on management tools built on the assumption that change will be linear and gradual to manage complex socioecological systems is a liability that is already costly and that has the potential to generate profoundly disruptive consequences for humans and their communities in the future. We are plagued by a problem of fit that is becoming more severe, despite growing sophistication regarding management tools and practices (Young 2002; Young et al. 2006b; Galaz et al. 2008). Or, to use Arild Vatn's phraseology, we run a growing risk of relying on policies that do not match the nature of the problems they are intended to address (Vatn 2012). As a result, achieving a transition to sustainability in the Anthropocene will require a substantial upgrade in the social capital available to meet needs for governance arising in complex systems (Brondisio, Ostrom, and Young 2009; Matson, Clark, and Andersson 2016).

In developing this argument and analyzing its implications, this chapter proceeds as follows. The first substantive section explores innovations in analytic tools available to address this problem of fit and asks how the use of new tools in a systematic and sustained manner would lead to management practices that differ from those characteristic of most current governance systems dealing with human-environment interactions. The next section introduces and examines the implications of the proposition that institutional arrangements themselves are complex systems characterized by a pronounced tendency to change in nonlinear and abrupt ways. The concluding section then seeks to distill some initial lessons from this account of complex and dynamic systems that support the development of what I describe as prudent practices regarding ways to recast resource and environmental regimes to achieve sustainable results in a world in which change can be nonlinear, abrupt, irreversible, and often nasty from a human perspective. This analysis together with the account of thresholds and trigger mechanisms in chapter 3 paves the way for an inquiry into innovative governance strategies in the chapters of part III.

Governing Complex Socioecological Systems Requires Innovation

Changes that are nonlinear, abrupt, irreversible, and surprising are common enough in both biophysical systems and socioeconomic systems. Phase or state changes affecting large ecosystems occur in many settings (Scheffer 2009). Much of the current concern about the fate of large marine ecosystems (LMEs), for instance, arises from the growing realization that these systems are susceptible to surprising system flips, a condition that calls into question familiar management practices governing consumptive uses of renewable resources (e.g., fish) and efforts to regulate the flow of pollutants into marine systems. Much the same is true of socioeconomic systems. What makes stock market crashes and business cycles so challenging, for example, is the occurrence of changes that are nonlinear and abrupt at the same time and that we seldom anticipate clearly enough to make a difference (Silver 2012).

The point of departure for this analysis, however, is the proposition that coupled or socioecological systems are more likely to give rise to changes that are, at one and the same time, nonlinear, abrupt, irreversible, and surprising than either biophysical systems or socioeconomic systems by themselves. This may seem paradoxical at first. After all, human systems are reflexive, and there is some reason to believe that humans will observe the early stages of major changes and take appropriate steps to prevent such changes from getting out of hand (Young et al. 2006b). But despite the effects of reflexivity, there is no evidence that humans have acquired the knowledge and the capacity to initiate and coordinate actions needed to manage changes that occur abruptly and on a large scale. Human reactions to changes in complex systems are just as likely to trigger positive or self-reinforcing feedback processes that increase the likelihood of explosions, cascades, and inflections as they are to prevent the occurrence of such nonlinear changes.

If this argument is correct, we will need a new generation of institutional arrangements or regimes to cope with the challenges of the Anthropocene. The fact that such changes will occur on a large scale means that we will need to address them through international and even global initiatives; uncoordinated national responses will not suffice. The abruptness of many of these changes will call into question current practices in which the formation of international or global regimes—much less their implementation—often takes years to decades and involves numerous false starts before progress is made (Young 1999). And the irreversibility of some of the largest and most nonlinear changes means that efforts to address the resultant problems through some procedure featuring trial and error will be risky. We will need to get our responses right at the outset or risk severe reductions in human welfare. It is easy to see that this will require substantial adjustments in the procedures we currently use to govern or manage human-environment relations at the macro level. But let me sharpen this issue and turn it into a question. What specific changes in institutional arrangements will be needed to succeed in governing for sustainability in a world of complex and dynamic socioecological systems? My initial answer to this question includes three related yet analytically distinct elements. I discuss them in this section under the headings of harnessing reflexivity, enhancing adaptability, and coping with uncertainty. With this analysis in hand, I will turn in the chapters of part III to a discussion of governance systems that differ fundamentally from mainstream regulatory arrangements and that can serve either as supplements or alternatives to regulatory arrangements in efforts to respond to various needs for governance in the Anthropocene.

Harnessing Reflexivity

Reflexive systems are those in which perceptions of recent or current trends give rise to expectations regarding future developments that trigger anticipatory responses. As this statement implies, we normally use the term reflexivity to refer to situations in which human actions are central to the dynamics of the relevant systems. The emphasis is on expectations and anticipatory responses framed in anthropogenic terms. It is possible to argue that other systems are reflexive, too, in the sense that non-human actors are capable of sensing patterns of change and taking actions designed to deal with expected rather than observed changes.[1] But in this discussion of governance involving complex and highly dynamic systems, the emphasis is on the capacity of human actors to identify trends, anticipate impending changes, and react in appropriate ways before the changes in question actually occur or at least before they get out of control.

It is striking how often human actors fail to identify trends and to anticipate nonlinear and abrupt changes that have far-reaching impacts on their welfare. Studies of the collapse of whole societies emphasize this phenomenon on a grand scale (Tainter 1998; Diamond 2005; Homer-Dixon 2006). But examples dealing with changes that are more limited in scope and closer to home are easy enough to locate. Few observers, much less actual policymakers, foresaw the collapse of cod stocks in the northwest Atlantic during the 1980s prior to their actual collapse in the early 1990s. Most observers were taken by surprise by the swift disintegration of the Soviet Union in the run-up to the final collapse at the end of 1991.

Even when human actors recognize prospective changes and take actions based on their expectations, the result may be to increase the likelihood of nonlinear, abrupt, and disruptive changes. We are all familiar with the idea of a run on the bank in which panicky actions on the part of sizable groups of individual actors ensure that widely feared events actually come to pass. This is the phenomenon captured in the idea of a self-fulfilling prophecy. Positive feedback processes that generate self-ful-filling prophecies can occur in many settings. The results range from fiscal crises that undermine otherwise viable economic systems to the self-gen-erating dynamics of arms races that prove immensely costly, do nothing to enhance the security of the participants, and may even precipitate wars that no one wants. Thus, the phenomenon of reflexivity may emerge as part of the problem rather than as part of the solution in efforts to devise governance systems capable of dealing with complex socioecological sys-tems in which nonlinear and irreversible changes are prominent realities.

Still, reflexivity can become an element of the effort to build effec-tive governance systems in such settings. The trick is to devise nega-tive feedback mechanisms that can kick in to alter the trajectory of the system before emerging trends cross a point of no return or a tipping point beyond which profound changes become inevitable. Familiar cases involve going on a diet to avoid the onset of obesity leading to fatal ill-nesses, and using monetary policy (e.g., cutting interest rates on the part of the Federal Reserve Bank in the US) to prevent economic downturns from triggering recessions or depressions. As these examples suggest, making use of reflexivity to address problems of governance is not easy. It is hard to forecast the consequences of gaining weight. Diets can produce undesirable consequences of their own, and they are often unsustainable. Cuts in interest rates on the part of central banks can trigger inflation, whether or not they succeed in their primary objective of stimulating eco-nomic activity.

Even so, there is much to be said for efforts to harness reflexivity as a means to maintain the fit between socioecological systems and rele-vant governance systems. Perhaps the most dramatic case in point today involves the looming threat of climate change. We do not know what level of greenhouse gases in the Earth's atmosphere will push the climate sys-tem past a tipping point, making major changes in this system unavoid-able and probably irreversible. Yet it is easy to see that recent trends in emissions of greenhouse gases are unsustainable (Alley 2000; Mayewski and White 2002). It is possible that limiting concentrations of carbon dioxide in the atmosphere to 450–500 ppm (parts per million) will suf-fice to avoid what we have come to refer to as "dangerous anthropogenic interference in the climate system." But many scientists (as well as envi-ronmental activists) now take the view that it would be safer to aim for

350 ppm (Rockström et al. 2009; Steffen et al. 2015). As our understanding of the climate system grows, pressure to take actions now to avoid costly impacts in the future is rising (Stern 2007).

Whatever its limitations, the development of the climate regime constitutes a significant effort to harness reflexivity. It seeks to mobilize actors who are preoccupied with more immediate problems—such as reducing extreme poverty, creating jobs, or controlling the cost of health care—to expend time, energy, and material resources on avoiding the occurrence of a future development that is hard for most people to grasp clearly and that may not generate costly impacts for some time to come. As our experience with integrated assessment modeling makes clear, moreover, there is profound uncertainty regarding both the probable costs of a failure to respond to changes in the Earth's climate system in a timely manner and the likely costs of taking effective steps to avoid the impacts of climate change (Stern 2007; Nordhaus 2008).[2]

As this example suggests, three key conditions govern the success of efforts to harness reflexivity to address large-scale problems associated with human-environment interactions. First is the issue of timing. How long should we wait before intervening, knowing not only that irreversible change is a distinct possibility but also that our understanding of the problem will improve over time and that our capacity to respond effectively is likely to grow? Is it accurate, for instance, to say that we have one or at most two decades left to make a determined effort to head off dramatic changes in the Earth's climate system? Second is the matter of collective action. How can we ensure that efforts to address climate change will take the form of concerted or coordinated actions? In cases like climate change where global actions are needed and incentives to free ride on the efforts of others are substantial, the challenge of coordinating responses looms large. Third is the problem of opportunity costs and avoiding unintended but potentially severe side effects of the actions we take. As those whose primary concern has been to fulfill the Millennium Development Goals know well, a desire on the part of wealthy developed countries to accord priority to the problem of climate change can divert attention and resources from efforts to address the problems of extreme poverty, inadequate sanitation, and substandard health care that are central concerns in the developing world (Young and Steffen 2009).

It follows that harnessing reflexivity is easier said than done. Still, this is a distinctive feature of human systems that can help to avoid the dangers arising when governance systems lack the nimbleness or agility to adapt by reacting quickly and decisively to changes in broader socioecological systems that are nonlinear, abrupt, or irreversible. To the extent that governance systems are unable to adjust rapidly, it makes sense to get a jump on problems by anticipating their occurrence and reacting in advance rather than waiting until crises are upon us before reacting.

Enhancing Adaptive Capacity

It takes no genius to see that success in governing complex socioecological systems subject to changes that are nonlinear, abrupt, irreversible, and surprising will require an ability to detect and interpret correctly the onset of such changes coupled with the flexibility needed to adjust the attributes of institutional arrangements to take into account the (anticipated or expected) consequences of the changes. But what does this mean in practice? Is there a danger that the arrangements required to deal with these challenges will necessitate costly compromises in terms of other conditions that are important to the achievement of success in regimes created to steer or manage human-environment interactions? If so, what can we do to limit the impact of such compromises without sacrificing the ability to create and maintain the required level of adaptability? This subsection addressees these questions with a focus on monitoring, early warning, and adaptive management.

Everyone will agree with the general proposition that effective monitoring is critical to any effort to steer or manage complex socioecological systems. The more these systems are prone to changes that are nonlinear, abrupt, and irreversible, moreover, the greater the need for high quality monitoring. But what exactly does this mean in practice? Knowing what to monitor may require both a sophisticated understanding of the nature of the socioecological system in question and the development of a set of indicators that can track the behavior of the system well and that are widely acknowledged to be credible and legitimate. Systems featuring negative feedback mechanisms that trigger countercyclical processes require less monitoring than those in which positive feedback mechanisms can lead to runaway processes. For some purposes a one-dimensional indicator (e.g., GDP per capita) may suffice. But an index that encompasses several dimensions (e.g., the UN Human Development Index) will prove more useful in other situations (UNDP 2006; UNDP 2007).

There are obvious questions concerning both the frequency of observations and the number of independent observations needed to obtain an accurate picture of the status of specific systems. Observations taken annually or even at multi-year intervals (e.g., the typical ten-year census cycle) may suffice for some purposes but not for others. A small number of observations may be sufficient in tracking homogeneous systems yet prove inadequate with respect to systems that are characterized by high levels of diversity. In dealing with marine systems, which are three-dimensional, fluid, and relatively opaque to many forms of observation, monitoring is a much more challenging task than it is in most terrestrial systems. Monitoring systems designed to anticipate abrupt changes (e.g., volcanic eruptions, tsunamis, hurricanes) will require much more frequent observations than monitoring systems in which change is slow, even though it may turn out to be nonlinear. Continuous monitoring in real time may be needed in some cases.

There is much to be said for enhancing monitoring capacity and building redundancy into observing systems. Yet resources are always limited, and monitoring can become an expensive proposition. So, we frequently face the following dilemma. The more we find ourselves dealing with changes that are nonlinear, abrupt, and irreversible, the greater our need for sophisticated and continuous monitoring. But such conditions also give rise to other needs (e.g., the development and maintenance of a rapid response capacity to cope with the impacts of abrupt events like earthquakes). The result is a need to consider tradeoffs or, in other words, to decide what to give up or forego in order to free resources to enhance monitoring capacity. Abstract or generic formulas that call for adding resources to monitoring until the marginal benefits from doing so just equal marginal costs will not get us far in such situations. In cases like the loss of biological diversity where we do not have certain knowledge about the existence of many—perhaps most—species, we do not even know what to monitor (Wilson 2003). On the other hand, when it comes to climate change, where it makes sense to think of the Earth's climate system as a single, integrated whole, the case for monitoring on a planetary scale is compelling.

What are the implications of these observations for specific governance systems like the climate regime or the regime for biological diversity? One response to this question is to design monitoring efforts to maximize returns on our investment. In dealing with biodiversity, for example, there is much to be said for focusing attention on the condition of keystone species or on mega-diverse regions or hot spots that are unusually rich in species. There is a similar argument for paying particular attention to areas that can function as the canary in the coal mine with regard to climate change (ACIA 2004). In all cases, we should be on the lookout for thresholds and trigger mechanisms. Such mechanisms produce results that are disproportionate to their role in causal clusters. Their value from the point of view of monitoring lies in the fact that they may help us to identify conditions leading to the onset of changes that are nonlinear in nature. Given the constraints arising from competing demands on available resources (economists would call them income effects), there is a good case for approaching this issue in investment terms, adopting strategies that are both focused and intensive.

These observations lead naturally to the proposition that we are in need of effective early warning systems. It makes sense to design monitoring systems in such a way as to provide as much warning as possible regarding approaching thresholds and the trigger mechanisms that can precipitate nonlinear and otherwise surprising changes. But whereas the case for monitoring is always persuasive, early warning becomes important under specific conditions. When changes are linear, gradual, and reversible, we will naturally want to track their trajectories. But there is no particular need to make a large investment in early warning. Even nonlinear changes may occur in slow motion (e.g., the destabilization of the Greenland ice sheet), giving those likely to be affected by state changes in relevant systems ample time to take note of the changes and to act in ways that eliminate, or at least alleviate, the negative impacts of nonlinear changes on human welfare. But early warning is of great importance in cases involving abrupt changes (e.g., earthquakes, hurricanes, floods, or volcanic eruptions). This leads to the following formula: Monitor changes in all cases, but focus on early warning only in those cases where abrupt changes are distinct possibilities and the events in question are likely to produce major impacts on human welfare.

Whether or not the case for early warning is compelling, institutional arrangements dealing with complex socioecological systems must have a capacity to adjust nimbly or agilely in the face of change. Analytically, we can distinguish among adaptive changes occurring at three distinct levels. First is the operational or micro level at which managers make adjustments that do not require changes in the prevailing rules of the game or approval on the part of governing bodies. Changes in the length of seasons (adjusting openings and closings) or alterations in total allowable catches in specific fisheries exemplify adjustments of this sort. At the next or meso level, adjustments typically involve the introduction of new rules or rights or shifts in the obligations of subjects, without altering the defining features of the regime itself. The case of the ozone regime in which phase out schedules can be accelerated without triggering the need for ratification on the part of member states (though the addition of new families of chemicals to those slated for phase out does require explicit consent) provides a prominent example (Parson 2003). At the third or macro level lie those situations involving the transformation of existing regimes or the creation of new ones in order to meet rising needs for governance. The replacement of traditional command-and-control regulations in the fisheries with arrangements relying heavily on incentive mechanisms in such forms as individual transferable quotas (ITQs), designated access privileges (DAPs), or tradable catch shares constitutes a prominent case in point. Even broader are proposals calling for the elimination of sector-specific arrangements (e.g., fisheries regimes) and their replacement with more integrated arrangements of the sort envisioned in some proposals for the introduction of place-based management (Crowder et al. 2006; Young et al. 2007).

The widely discussed but somewhat vague idea of adaptive management comes into focus here (Lee 1993). Adaptive management applies to the first level and probably the second level identified in the previous paragraph. It makes sense to experiment with the use of specific policy instruments, such as no-take zones and gear restrictions in the fisheries or alternative phaseout schedules in the case of ozone-depleting substances. But when it comes to more transformative changes like the shift from systems focused on obtaining maximum sustainable yields (MSY) from fish stocks to ecosystem-based management (EBM) in ocean governance or the creation of quasi-markets designed to reduce atmospheric pollution, it is probably better to speak of societal learning (Social Learning Group 2001). At this level, we are replacing existing arrangements with something qualitatively different rather than adjusting arrangements already in place in the light of new information about the problem to be solved or the efficacy of specific policy instruments.

At each level, there is a need for sufficient flexibility to adjust or even to replace existing arrangements in the face of changes that are nonlinear, abrupt, and irreversible. Yet this can become a slippery slope with regard to the effectiveness of governance systems. Any arrangement that changes in response to every little shift in the (perceived) nature of the problem or the behavior of the subjects cannot be effective in governing human-environment interactions. Systems that are extremely difficult or even impossible to adjust in the light of changing circumstances, on the other hand, will become brittle and ultimately collapse as circumstances change. The result is a challenge to those charged with (re)designing management systems. In this context, the gap between rules on paper and rules in use may be large.

The Antarctic Treaty of 1959, to take a specific example, allows for the convening of a review conference at the request of any contracting party at any time after the treaty has been in force for thirty years, or from 1991 onward since the treaty entered into force in 1961. But the parties have never activated this provision, despite the occurrence of matters that could and perhaps should be revisited in the light of experience with this regime. The problem arises from the fact that many members of the regime believe that holding a review conference would open a Pandora's box of proposed changes and that the overall results of such a conference would constitute a step backward rather than a step forward. Consequently, they prefer to make adjustments in a piecemeal manner rather than engaging in a more comprehensive review process.

The levels of change identified in the previous paragraphs are relevant in this context. It makes sense to increase the stringency of the procedures governing adjustments as we move from the micro level to the macro level. Experiments designed to fine tune existing policy instruments (e.g., the details of the cap-and-trade system applying to emissions of sulfur dioxide in the United States) make a lot of sense. But we are understandably cautious when it comes to major adjustments in constitutive agreements or even proposals to replace them with new and substantially different arrangements. Consider the differences among adjusting existing regulations, passing new laws, and adopting amendments to the Constitution of the United States as a case in point. Under most circumstances, it makes sense to provide administrative agencies with the discretionary authority needed to adjust regulations in the face of change, while at the same time adopting rules and procedures that make it much harder to enact amendments to the constitution itself.

Another point worth noting in this discussion of adaptive capacity has to do with forms of change. While some changes are formalized and codified through constitutional amendments or legally binding agreements, others involve informal adjustments to day-to-day practices. The challenge here is to match forms of change with the circumstances at hand, rather than asking whether one form of change is preferable to the others in some general sense. Informal changes have the virtues of being easier to adopt, quicker to implement, and less controversial to adjust. But formal or legally binding changes have greater normative force and put more pressure on subjects to comply with their requirements or adjust their behavior to meet the requirements of new systems of rights and rules. In cases where changes are nonlinear and especially abrupt, relying on informal adjustments is often essential. Changes in constitutions or other constitutive documents involve deliberative processes that under most conditions cannot be rushed. There may be good reasons to update constitutions from time to time to formalize major developments in the de facto character of governance systems as well as to adjust to socioeconomic changes. But when changes are nonlinear, abrupt, irreversible, and surprising, the case for relying on informal changes—at least in the short run—is compelling.

Coping with Uncertainty

No matter how much effort we devote to analysis and other forms of preparatory work, there is no way to avoid uncertainty in governing human-environment interactions involving complex socioecological systems. Uncertainty involves unknowns occurring at several levels. Ordinary uncertainty is a matter of our inability to provide clear-cut answers to questions of the following sort. How many distinct species are there on the planet? For any given species, how low can the population go before it becomes a relic or ghost species? At what level will concentrations of greenhouse gases in the atmosphere give rise to dangerous anthropogenic interference in the climate system? What is the probability that highly disruptive rapid climate change events (RCCEs) will occur during the next few decades?

Beyond this lies what we can call second-order uncertainty, a higher level of uncertainty that has to do with unknown unknowns. Are there as yet undetected forces that can help to explain why the Earth's climate system has behaved in a benign manner over the last 10,000 years and that will determine the likelihood that this situation will continue into the future? Will globalization affect large-scale socioecological systems in ways that we have yet to conceptualize, much less to analyze in a systematic manner?

Recent thinking about biophysical systems has emphasized a holistic approach, a perspective that highlights dynamism, nonlinear processes, state changes, and the importance of emergent properties. Similar comments are in order regarding socioeconomic systems. As the stories of numerous failed states attest, these systems can experience state changes that take even attentive observers by surprise. But where uncertainty becomes particularly wide and deep is in efforts to understand the dynamics of socioecological systems. Here we face not only uncertainty associated with the behavior of biophysical systems and socioeconomic systems on their own, but also the higher-order uncertainty that arises in conjunction with coupled systems in which biophysical and human forces become interactive drivers and emergent properties are particularly prominent. This variety of uncertainty is destined to grow in importance as we move deeper into the Anthropocene or, in other words, into an era of human-dominated systems in which it is pointless to attempt to analyze either the biophysical or the anthropogenic forces at work in terms of models that set one or the other of these types of forces aside as exogenous factors.

There is much more to be said about specific types of uncertainty and about the prospects for reducing or even eliminating uncertainty regarding some issues (Kahneman and Tversky 2000). But the important point here is that regimes governing human-environment interactions in complex systems must be able to function effectively in situations featuring more or less severe uncertainties. Among the strategies available for coping with these uncertainties are the adoption of precautionary measures including insurance schemes, the development of heuristics or rules of thumb, and the conduct of targeted research.

One way to respond to uncertainty is to adopt a precautionary approach. In other issue domains (e.g., national security), we routinely base policy choices on some form of worst-case analysis. We assume the worst about both the capabilities and the intentions of potential opponents and construct our defense forces with this assumption in mind. Following this logic, we routinely allocate hundreds of billions of dollars—currently around 3.5 percent of GDP in the U.S.—to defense spending every year. This is not to say that we should adopt a worst-case approach to potential problems arising in human-environment interactions; it is worth asking whether extreme forms of worst-case thinking have done more harm than good even in the case of national security. Nonetheless, it is striking how reluctant we are to consider and act on increasingly plausible scenarios about threats associated with RCCEs. A precautionary approach in this case would center on making deep cuts in emissions of greenhouse gases justified by the possibility of RCCEs occurring, even though we are not able to forecast either the timing or the consequences of such events with precision. Calculations of the costs of adopting such an approach in the case of climate suggest that taking action would be expensive (costing perhaps 1–2 percent of GDP) but not nearly as expensive as the measures we take year after year in the name of national security.

A related but distinct response to uncertain events centers on the development of insurance schemes. The purpose of insurance—in such forms as life insurance and homeowners' insurance—is to avoid ruination in the face of low probability but catastrophic events, such as a premature death or the destruction of a home by fire (Legget 2001). Useful as these devices are in their normal applications, they have major limitations as procedures for dealing with problems like climate change that grow out of human-environment interactions under conditions of uncertainty. It is virtually impossible to approach threats like climate change in actuarial terms. We are dealing here with events that are to a large extent one of a kind. The best we can do under the circumstances is to employ subjective probabilities, a procedure that is often contested and requires a willingness to address normative issues, such as whether or how to discount the significance of events that may occur at some time in the future or the welfare of members of future generations. The strategy of taking out insurance also gives rise to the problem of moral hazard. To the extent that they are confident that either private or public insurance systems will indemnify their losses, individuals may take greater risks than would otherwise be the case. We see this clearly in such contexts as the destruction wrought by the impact of hurricanes on homes located on barrier islands or the impact of wild fires on homes located in areas prone to fire. So long as the government is willing to designate hard hit areas as disaster zones and to provide financial aid on generous terms to victims of such destructive events, owners of homes in areas prone to flooding or burning are more likely to rebuild in the same place instead of relocating than would be the case in the absence of public assistance.

Another response to the uncertainties associated with human-environment interactions is to rely on various types of heuristics or rules of thumb (Tversky and Kahneman 1974). The use of heuristics is widespread in situations involving uncertainty, which is to say in most situations that find their way onto public policy agendas. There is simply no way to avoid making decisions about a raft of issues (e.g., national defense, macroeconomic policy, health care, public education) in which difficulties of measurement abound and uncertainties about the effectiveness of specific policies loom large. One way to respond to this situation is to make use of rules of thumb to bridge the gap. Such rules may cover a wide range of situations—from expectations about the extent to which subjects will comply with regulatory measures to expectations about the costs of achieving various environmental goals. We may hope that such rules reflect some sort of intuitive reasoning, even in cases where it is impossible to establish causal connections. Rules of thumb can reflect best practices in the sense that they arise from a deliberate examination of a number of cases that resemble one another in significant ways (e.g., they all require finding ways to avoid the tragedy of the commons or to deal with externalities that impinge on the welfare of others). Such rules always rest on subjective calculations or qualitative judgments; applying them to specific situations can produce results that run counter to the goals of policymakers. Still, the case for developing such rules is strong. Framing best practices in the form of recommendations relevant to current issues is a better response to uncertainty than a procedure reflecting nothing more than ad hoc or "seat-of-the-pants" judgments.

It is naïve to suppose that research can eliminate the uncertainties that plague efforts to make good choices regarding issues that arise in human-environment interactions. Nonetheless, applied research may prove helpful to policymakers responsible for making such choices. One prominent case involves the conduct of scientific assessments of the sort we associate with the work of the Intergovernmental Panel on Climate Change (IPCC) and the Millennium Ecosystem Assessment (MEA) (IPCC 2007; IPCC 2014; MEA 2005). These projects strive to synthesize all available knowledge bearing on the issues at hand. They do not feature the conduct of new research, and they do not eliminate uncertainty surrounding major issues like the trajectory of climate change or the loss of biological diversity. Yet it is widely acknowledged that these projects have proven helpful to policymakers despite—or perhaps because of—the uncertainties surrounding such issues (Mitchell et al. 2006). The IPCC's assessment reports do not tell us how high concentrations of greenhouse gases in the atmosphere can go without triggering "dangerous anthropogenic interference in the climate system." Those responsible for IPCC assessments are careful to give a range of estimates regarding such matters and to stress that these estimates should not be construed as predictions (Bolin 1997). Even so, the work of the panel has generated useful results on issues ranging from the atmospheric processes that give rise to climate change to the probable results of various proposals for carbon sequestration or carbon capture and storage.

Simulations of the sort associated with integrated assessment modeling constitute another useful research endeavor. Such efforts join together biophysical components (e.g., climate models) with socioeconomic components (e.g., macroeconomic models) to explore the interactions characteristic of coupled and complex systems. The simulations arising from these exercises are not predictive in any meaningful sense. But they are useful in exploring the dynamics of complex systems and understanding the idea of emergent properties. They allow us to analyze the sensitivity of some components of the system to relatively small changes in other components, increasing awareness of situations featuring cascades of change and surprising outcomes in the process. A particularly interesting case in point centers on the development of models designed to explore the effects of alternative discount rates in comparing the costs to society arising from the impacts of projected changes in the climate system with the costs of launching sustained efforts to lower the probability that such changes will occur (Stern 2007; Nordhaus 2008).

There is also a role for scenarios in efforts to think productively about situations featuring high levels of uncertainty. Scenarios are narratives or stories about plausible future developments in complex systems. They are not simply projections of current trends, and they are not meant to have predictive value. Scenarios are useful in encouraging analysts to think about situations that reflect dramatic departures from business as usual and structuring efforts to evaluate options that become relevant in such situations. Good scenarios direct attention to situations that are realistic enough to be plausible, yet different enough from current conditions to require analysts and policymakers to consider dramatic shifts in prevailing governance systems. Their role is to push us toward considering nonlinear changes but in a controlled and rigorous fashion.

Institutions Are Complex Systems, Too

The previous section addressed the nature of complex socioecological systems and explored the features of institutional arrangements needed to steer or guide these complex systems toward socially desirable outcomes. The basic message is clear. The combination of highly dynamic socioecological systems and high levels of uncertainty regarding the behavior of these systems calls for the creation and operation of institutional arrangements that differ in significant ways from the regimes we normally think of in our efforts to address issues of sustainability. But there is another source of complexity we need to consider in this context as well. Institutional arrangements are systems, too, and there are good reasons to regard them as complex systems just like the socioecological systems they are created to govern. This means we must be prepared for failure in our efforts to govern or manage specific human-environment interactions. But we also need to redouble our efforts to understand the challenge of governance in this setting and to devise arrangements that are both adjustable in the face of change and able to tolerate high levels of uncertainty.

Consider, again, the issues of linearity, speed, and reversibility as features of change. Nonlinear change is a common occurrence with regard to institutional arrangements. There is a sense in which the creation of a new regime or a fundamental change in an existing regime constitutes a form of planned or intentional nonlinearity. Given the complications associated with implementation or the transition from paper to practice with regard to institutional arrangements, such changes may be somewhat less dramatic on the ground than they appear to be on paper. Nevertheless, the creation of a governance system like the regime designed to protect the stratospheric ozone layer through the adoption of the 1985 Vienna Convention for the Protection of the Ozone Layer and the 1987 Montreal Protocol on Substances that Deplete the Ozone Layer produces a distinct break or discontinuity with the situation prevailing prior to the adoption of the relevant agreements (Parson 2003). Although the mechanism was less formal in the sense that it did not involve altering legally binding arrangements, we can say also that the adoption in 1982 of a moratorium on the harvesting of great whales gave rise to a nonlinear shift in the character of the regime established under the terms of the 1946 International Convention for the Regulation of Whaling (Friedheim 2001). Similar breaks occur at other levels of social organization as well. The decision in the United States to reduce emissions of sulfur dioxide by roughly 50 percent and to create a cap-and-trade system to achieve this goal under the provisions of the Clean Air Act Amendments of 1990 constitutes a nonlinear change at the national level (Bryner 1995).

Institutional arrangements can also pass thresholds and experience trigger mechanisms that precipitate rapid changes of an unintended nature. Under normal circumstances, regimes are remarkably path dependent, sometimes remaining in place long after the problems they address change substantially and mismatches between problems and institutional arrangements become prominent. Yet abrupt changes in institutions are also common; they often take by surprise even those who are closely involved in their management. We are all familiar with such cases at the societal level; the collapse of the Soviet Union at the end of 1991 is a striking example. But similar changes occur from time to time with regard to environmental and resource regimes. The regime designed to manage the harvesting of northern fur seals, which had a history stretching back to the early years of the twentieth century and which had a favorable track record, collapsed like a house of cards in 1984–1985, when the US Senate refused to approve what previously would have been a routine extension of its provisions (Mirovitskaya, Clark, and Purver 1993). The international regime governing exchange rates and related monetary issues experienced a sea change in 1971, when the US decided that it would no longer treat the dollar as convertible to gold. Nor is abrupt change confined to the collapse or dramatic restructuring of regimes. Efforts to create regimes that seem stuck and unlikely to bear fruit can experience breakthroughs when diplomatic or legislative logjams are broken. An interesting case in point is the regime governing the administration of the Svalbard Archipelago. The provisions of the 1920 Treaty of Spitsbergen, adopted almost as an afterthought in the context of the peace settlement following World War I, would have been impossible to adopt in the political environment prevailing in the pre-war period (Singh and Saguirian 1993).

Similar remarks are in order regarding the magnitude of change, at least relative to the scale and scope of issues that environmental and resource regimes address. It is hard to imagine a more fundamental change than the shift from a regime designed to manage consumptive uses of whales sustainably to an arrangement based on the idea of preservation and intended to terminate all killing of these animals. The decision of the US to mandate sharp reductions in emissions of sulfur dioxide as a means of addressing the problem of acid rain also deserves to be treated as a fundamental change. It represented a sharp break from a system in which emitters were permitted to use the atmospheric commons as a repository for toxic wastes free of charge, to a system in which this use of the atmosphere is acknowledged to amount to the use of a scarce resource that should be limited and priced accordingly. Institutional changes that are incremental or so modest that they are hard to identify are also common. This is especially true regarding contested issues where changes are clothed in language that is so opaque that it is difficult to determine whether or not they will amount to anything over time. Overall, we need to recognize that environmental and resource regimes are not static arrangements; change of one sort or another is a continuous process in such settings.

Most institutional changes are irreversible in fact if not in principle. Like socioecological systems that have been disturbed, regimes that are subjected to more or less severe stresses seldom settle back into patterns that replicate the status quo ante, even after a crisis or a significant perturbation subsides. It is normal to add new provisions to a regime designed to avoid or to cope with the forces giving rise to a crisis. Whether or not they prove effective in specific cases, the introduction of limited-entry schemes to stabilize fisheries suffering from severe depletions and overinvestment in boats and fishing gear reflects such reasoning (Costello et al. 2008). But disturbances can also lead to other types of change, such as the dramatic extension of coastal state jurisdiction reflected in the creation of Exclusive Economic Zones formalized in the 1982 UN Convention on the Law of the Sea (Iudicello et al. 1999). There is no inherent reason why institutional changes should be irreversible; it is possible to imagine tentative or experimental changes that are allowed to expire or are rolled back in favor of the status quo ante. Nonetheless, it is hard to avoid the conclusion that institutional change is apt to become a one-way street. Institutional reforms may fail, and this may generate incentives to move on to yet another effort at restructuring prevailing systems of rights and rules. But a simple reversion to the status quo ante is seldom feasible.

It would be wrong to overdo these arguments. Many—perhaps most—institutional arrangements do have staying power. In fact, regimes often prove sticky and resistant to change, even when they are no longer well matched to the problems at hand (Young 2002). But the important point is this: not only are socioecological systems complex and given to changes that are nonlinear, abrupt, and irreversible; the institutional arrangements we create to govern human-environment interactions in such settings are also complex systems that exhibit similar patterns of change. The implication of this is clear but challenging: we are operating in a world of flux and uncertainty (Wilson 2006). Our goal is to match governance systems to socioecological systems to avoid or minimize the problem of fit. But both sides of this equation are dynamic, so that it is always risky to create institutional arrangements and then to assume that there is no need to monitor the resultant interactions and to make appropriate adjustments on a continuous basis.

So, How Should We Proceed?

What does this analysis tell us about how to adjust our normal thinking regarding the creation and operation of governance systems as we strive to achieve sustainability in the Anthropocene? There is no way to do justice to this broad question in a few paragraphs. Still, I think it is feasible to identify some prudent practices that should inform our efforts to steer human-environment interactions in the light of the argument I have sketched in this chapter. For starters, I suggest we focus on (i) making good use of various forms of simulation and scenario development to enhance our understanding of the dynamics of complex systems, (ii) monitoring the behavior of these interactive systems closely and establishing procedures that facilitate adjustment to changing circumstances, (iii) drawing a clear distinction between basic institutional elements that must be protected and operating rules that are subject to change, and (iv) constructing firewalls and building redundancy into institutional arrangements to minimize the likelihood of systemic collapse.

Prudent Practice 1: Take full advantage of opportunities to improve knowledge through the use of sensitivity analysis, simulations, scenario development, and other procedures that allow us to gain insight into the dynamics and emergent properties of complex and indeterminate systems. There is much to be said for devoting significant time and resources to the use of simulations and related procedures to gain insights regarding the likely results of institutional innovations in efforts to manage complex systems. Simulations do not allow us to make predictions regarding the behavior of real-world systems; they cannot provide a basis by themselves for selecting rules and decision-making procedures to deal with specific situations. But complex systems—both of the socioecological sort and of the institutional variety—behave in ways that generate emergent properties that are counterintuitive and that often take us by surprise. Even simple simulations, like the game known as Fishbanks, can generate insights and understanding regarding the consequences of human uses of common-pool resources that are otherwise hard for most users to grasp.[3] Similar remarks are in order regarding the use of scenarios to direct attention to the dynamics of complex systems. This first prudent practice, then, calls for concerted efforts to use simulations and scenarios in order to enhance awareness of the range of developments that may occur in complex systems and that managers should be aware of in making decisions about the use of specific policy instruments. The fact that both institutional arrangements and socioecological systems are complex and dynamic simply reinforces the relevance of this practice.

Prudent Practice 2: Monitor closely both the behavior of socioecological systems and the performance of institutional arrangements and devise procedures that make it easy to make adjustments or mid-course corrections to address the problem of fit. Not surprisingly, there is no alternative in dealing with situations of the type under consideration in this chapter to devoting the resources required to monitor the behavior of socioecological systems and the performance of environmental and resource regimes on a continuous basis. In cases where abrupt and potentially nasty changes (e.g., RCCEs in the case of climate change) are a real concern, the need for monitoring will be especially acute. But whenever we are dealing with changes that are likely to be nonlinear, abrupt, irreversible, and surprising, the costs of failing to monitor continuously are likely to be high. Our experience in this realm is full of catastrophic but avoidable changes, such as the American Dust Bowl crisis of the 1930s, the collapse of the Northwest Atlantic cod fishery in the 1980s–1990s, the destruction in New Orleans caused by hurricane Katrina in 2005, and the impacts of climate change occurring in the Arctic today. In each case, the failure has arisen in part from an inability to understand changes in socioecological systems combined with an unwillingness to introduce significant changes in prevailing institutional arrangements. Even when the prospect of nonlinear and abrupt change becomes evident, path dependence often thwarts efforts to craft timely and appropriate responses. The second prudent practice, then, calls not only for a dedication to continuous monitoring but also for the development of procedures that make it possible to respond to changing circumstances nimbly or agilely and without becoming bogged down in the political and bureaucratic complications that commonly delay processes of institutional reform.

Prudent Practice 3: Draw a clear distinction between the core elements or constitutive features of governance systems and the operating rules or policy instruments used to implement them. The point of this practice is to draw a distinction between foundational (typically normatively grounded) commitments and policy instruments and to spell out a rationale for treating them differently. In the realm of ocean governance, for instance, we can make a commitment to use ecosystem-based management and provide meaningful opportunities for stakeholder involvement, without deciding in advance whether to create certain types of marine protected areas or to rely on a particular sort of limited-entry system regulating harvesting of living resources. In the case of climate change, we can commit ourselves to making specified cuts in greenhouse gas emissions over a period of time, without making fixed choices regarding the use of carbon taxes, cap-and-trade arrangements, or some other regulatory mechanism to get there. This third prudent practice can succeed only if those responsible for the administration of the resultant regimes are genuinely committed to pursuing a common goal and are in possession of the resources needed to do so. This is a critical condition; it is easy to imagine circumstances in which interest groups of one sort or another will endeavor to manipulate or corrupt such arrangements to serve their own ends. But there is no substitute for the combination of commitment to foundational principles and flexibility with regard to the selection and adjustment of policy instruments. It is preferable to work hard to hold the line against corruption than to adopt a policy of relying on rigid institutional arrangements in a world of complex and dynamic systems.

Prudent Practice 4: In cases where adaptive capacity is limited, incorporate firewalls and redundancy into institutional arrangements to minimize the likelihood of abrupt collapse at the systemic level. One way to mitigate the dangers associated with the occurrence of nonlinear, abrupt, irreversible, and surprising change is to compartmentalize institutional arrangements to avoid cascading processes leading to systemic collapse, or to introduce redundancy so that governance systems can remain effective even when some of their elements are degraded or destroyed. Of course, this strategy will incur costs. As the literature on the currently popular concepts of multi-level governance and polycentric governance makes clear, coordination can be hard to achieve under such arrangements, and there is a tendency for such systems to privilege the interests of some stakeholders over the interests of others. At a minimum, this response to the dangers of nonlinear, abrupt, irreversible, and surprising change will involve significant opportunity costs. This suggests that the prescription embedded in this principle will be most relevant when the likely changes are large scale and the probability of passing critical thresholds is either high or unknown. A current issue that fits this description is the problem of climate change.

A Final Note

The articulation of these prudent practices is only a first step in learning to govern for sustainability in a world of complex socioecological systems. Efforts to manage human-dominated ecosystems will produce failures as well as successes. But the fact that human behavior is reflexive offers some basis for hope. Reflexivity can make matters worse, as the familiar metaphor of a race to the bottom suggests. Yet reflexivity can also help us to address problems in a timely manner, providing the impetus to innovate when things go wrong and to learn from experience in assessing the results of institutional innovations. Some such approach will be essential if we are to find ways to come to terms with the dynamics of the complex socioecological systems and institutional arrangements that are defining features of the Anthropocene and, in the process, to succeed in navigating the sustainability transition. With these observations in hand, I turn in the chapters of part III to a consideration of governance systems that differ in important ways from the mainstream regulatory approaches that have dominated much of our thinking about governance in recent times.

III

New Perspectives on Governance

5
Goal-Setting as a Governance Strategy

Introduction

The challenge of meeting needs for governance has emerged as a central concern in many social settings, not least in international society where we are confronted with the complexities of devising integrated responses to the interactive biophysical and socioeconomic forces affecting the pursuit of sustainability in the absence of a system-wide government to take responsibility for meeting these needs. In considering ways to address this challenge, most of us think first of regulatory arrangements, emphasizing the development of rules and focusing on issues relating to the implementation of the rules and on procedures that are useful in eliciting compliance on the part of those subject to the rules (Chayes and Chayes 1995). But if we think of governance in generic terms as a social function centered on steering individuals or groups toward desired outcomes, we can broaden the scope of our thinking by considering goal-setting and efforts to meet targets associated with key goals as a distinct strategy for fulfilling needs for governance in a variety of settings.

In this chapter, I explore the nature of goal-setting as a governance strategy, analyzing conditions under which goal-setting can prove effective as a steering mechanism, considering the strengths and weaknesses of goal-setting, and assessing the usefulness of goal-setting in efforts to govern complex systems of the sort that are an increasingly important feature of the Anthropocene. I draw on efforts under the auspices of the United Nations to launch and implement the Millennium Development Goals (2000–2015) and the Sustainable Development Goals (2016–2030) as sources of examples, since these initiatives constitute the most prominent contemporary examples of goal-setting on a global scale. But my objective is to engage in a general assessment of the uses of goal-setting as a means of coping with the challenges of governing complex systems. The argument of this chapter is empirical rather than normative or prescriptive in character. I make no effort to pass judgment on the relative merits of rule-making and goal-setting as distinct governance strategies. Rather, I seek to shed light on goal-setting as a strategy for responding to needs for governance that has received less attention than rule-making among those who think about governance at the international or global level and specifically among those concerned with the pursuit of sustainability in the Anthropocene.

In the body of the chapter, I proceed as follows. The first substantive section explores the basic character of goal-setting as a governance strategy and differentiates it from the more familiar idea of rule-making. The following section comments on circumstances under which it may be expedient to join goal-setting and rule-making to form integrated governance systems. The next section identifies a number of pitfalls that plague efforts to employ goal-setting to good effect in large-scale settings like international society. The chapter then turns to a more general account of the determinants of the effectiveness of goal-setting as a governance strategy. This sets the stage for some specific observations about the effectiveness of goal-setting under the conditions characteristic of the Anthropocene. The penultimate section discusses procedures or mechanisms that may prove helpful to those desiring to enhance the effectiveness of goal-setting in a variety of settings. The final substantive section illustrates how the strategy of goal-setting works in practice through a discussion of efforts leading to the establishment and implementation of the Millennium Development Goals (MDGs) and the Sustainable Development Goals (SDGs).

Goal-Setting as a Governance Strategy

Goal-setting seeks to steer behavior by (i) establishing priorities to be used in allocating both attention and scarce resources among competing objectives, (ii) galvanizing the efforts of those assigned to work toward attaining the goals associated with the resultant priorities, (iii) identifying targets and providing yardsticks or benchmarks to be used in tracking progress toward achieving the goals, and (iv) combating the tendency for short-term desires and impulses to distract the attention or resources of those assigned to the work of goal attainment. Goal-setting thus differs from rule-making, a governance strategy that seeks to guide the behavior of key actors by articulating prescriptive rules (and implementing regulations) spelling out requirements and prohibitions and devising compliance mechanisms whose purpose is to induce actors to adjust their behavior accordingly.

To make these somewhat abstract ideas concrete, think of the capital campaigns that universities, hospitals, libraries, public radio stations, and various charitable organizations launch from time to time as a source of illustrative examples. The usual procedure is to set a concrete goal defined in monetary terms, lay out some plans regarding the use of the money to be mobilized, specify a target date for meeting the goal, publicize the campaign vigorously, and create a highly visible system for tracking progress. The idea is to activate known supporters, while at the same time identifying and signing up new supporters. Such campaigns not only serve to focus the efforts of regular staff members and to provide guidance for enthusiastic volunteers; they also serve to mobilize resources and set the relevant organizations on paths that are likely to shape their programmatic development for years to come. The effectiveness of capital campaigns is by no means assured. There are highly paid experts whose business it is to advise organizations on when to launch a capital campaign and how to decide on an appropriate goal. But well-planned and well-timed campaigns regularly prove successful in meeting their goals. Some even exceed their goals, an outcome that allows leaders to make credible claims regarding their prowess as fundraisers.

This illustration is helpful as a means of revealing the defining features of goal-setting as a steering mechanism. Taking this conception of goal-setting as a point of departure, we can identify relatively well-defined examples of the use of this governance strategy at the national and international levels. Prominent cases at the national level include the reconfiguration of the American economy during 1942–1943 to fulfill the goal of turning the United States into an "arsenal of democracy" in the fight against the Axis powers (Goodwin 2009) and, perhaps even more clearly, the work of the American Apollo Project launched by the Kennedy Administration and committed to putting a man on the moon by the end of the decade of the 1960s. Undoubtedly, the most prominent recent example of goal-setting at the international level features the development and implementation of the Millennium Development Goals launched in 2000 through the adoption of the UN General Assembly's Millennium Declaration and granted a fifteen year lifespan starting in 2001 and running through 2015 (UN General Assembly 2000). The 2015 Paris Agreement on climate with its emphasis on sharpening the operationalization of the goal and its focus on voluntary pledges in the form of Intended Nationally Determined Contributions (INDCs) reflects a growing interest in goal-setting as a way forward in addressing challenges to sustainability on a global scale.

Although the particulars of goal-setting vary greatly from one situation to another, all efforts to make use of goal-setting as a governance strategy share three distinctive features. Goal-setting requires, to begin with, an ability to establish well-defined priorities and to cast them in the language of explicit goals. The whole point of goal-setting is to single out a limited number (sometimes just one) of concerns and to accord them priority in the allocation of scarce resources including staff time and political capital. Once the goals are established, efforts to attain goals normally proceed in a campaign mode. The essential idea is to galvanize attention and to mobilize resources to make a sustained push to achieve measurable results within a fixed timeframe. The goal of the Apollo Project to put a man on the moon within a decade provides a striking but representative example. In addition, goal-setting requires an effort to devise clearcut metrics to track progress over time. MDG 1, aimed at halving the number of people living on less than US $1 a day by 2015, provides a good example, though collecting the actual data needed to track a metric of this sort may prove difficult under real world conditions. Not only do tracking mechanisms play a key role in measuring progress toward goal attainment; they are also useful as a means of urging all those involved to redouble their efforts to fulfill the goal by the predetermined cutoff date.

The essential premise of goal-setting as a governance strategy consequently differs from the premise underlying rule-making. Whereas rule-making features the formulation of behavioral prescriptions (e.g., requirements and prohibitions) and directs attention to matters of compliance and enforcement, goal-setting features the articulation of aspirations and directs attention to procedures for generating enthusiasm among supporters and maximizing the dedication needed to sustain the effort required to reach relatively well-defined targets. Moreover, whereas goal-setting normally features the mounting of a campaign designed to attain goals within a specified timeframe, rule-making features the articulation of behavioral prescriptions expected to remain in place indefinitely.

Linking Goal-Setting and Rule-Making

Both goal-setting and rule-making can and often do operate on their own as distinct strategies for addressing governance needs. Consider for illustrative purposes the contrast between the goal-setting strategy reflected in the MDGs and the rule-making approach embodied in many regimes aimed at steering the activities of those engaged in harvesting living or renewable resources (e.g., fisheries regimes). The MDGs called on parties to mount a campaign to achieve goals like eradicating extreme poverty and hunger, reducing childhood mortality, and combating HIV/AIDS, malaria, and other infectious diseases, all within a timeframe extending through 2015. Regimes governing the use of renewable resources, by contrast, typically rely on rules dealing with such things as permits or licenses required to engage in harvesting the resources, quotas, open and closed seasons, restrictions on gear types, the treatment of by-catches, and so forth. Often, there is little overlap regarding the mechanisms used in the two types of governance systems.

Nevertheless, the two strategies are not mutually exclusive. They may prove complementary in specific situations. Even more to the point, goal-setting and rule-making may become elements in integrated governance systems. Consider as examples the goal of avoiding "dangerous anthropogenic interference with the climate system" articulated in the UN Framework Convention on Climate Change as the overall objective of the climate regime (UNFCCC 1992, Art. 2) and the goal of achieving maximum sustainable yields (MSY) or even optimum sustainable yields (OSY) embedded in regimes for fisheries or marine mammals.[1] In such cases, regulatory arrangements are introduced as means of steering behavior toward the achievement of explicit goals. Situations of this sort are common and deserve systematic attention in any comprehensive account of governance strategies. But the principal focus of this chapter is on the use of goal-setting as a distinct governance strategy in situations featuring free-standing goals like the MDGs and now the SDGs.

Common Pitfalls of Goal-Setting at the International Level

Goal-setting is intuitively appealing both because we all have experience with setting goals in our private lives and because most of us feel that the pursuit of such personal goals has played or can play an important role as a means of focusing and guiding our efforts or, in other words, as a form of self-governance. But can we generalize from our personal experiences with goal-setting employed as a strategy for achieving self-governance to the use of goal-setting at the societal level treated as a strategy for achieving collective governance, especially at the level of international society?

In addressing this question, a number of differences between self-governance and collective governance come into focus that are apt to pose problems for the use of goal-setting in large-scale social settings like international society. Consider the following issues as illustrations of this proposition.

Setting the goals—Setting priorities is a difficult process under the best of circumstances. Individuals often experience inner conflict in the effort to establish personal priorities. But when goal-setting is a collective endeavor involving some form of negotiation or consensus-building among a relatively large number of self-interested actors (e.g., states in international society), there is a danger that (i) the group will end up with too many goals to be helpful in establishing priorities and allocating scarce resources, (ii) the goals chosen will be framed in vague terms that are hard to operationalize much less to monitor, or (iii) the individual goals included in a package will be incompatible or even contradictory?

Tracking progress—It is easy to track progress in the case of a capital campaign where the goal is stated in monetary terms. Devices like a "barometer" showing the percentage or proportion of the total raised at any given time are familiar tracking methods used in this context. But many international goals are framed in terms that are hard to measure and that result in the use of procedures that are inadequate or even misleading. In tracking progress toward improving human welfare, for example, there is a danger of overemphasizing operational criteria (e.g., GDP per capita), underemphasizing values that are critical but harder to measure (e.g., broader measures of well-being or human development), or ending up with a high level of uncertainty or disagreement about the issue of tracking progress toward goal attainment (e.g., the operational meaning of a concept like sustainable development) (Karabell 2014)?

Behavioral mechanisms—The incentives/pressures to stick with goals may be inadequate to steer behavior at the international level, especially when goal attainment is a slow process or the going gets tough due to competing demands or shortages of resources during periods of economic, political, or social stress. The mechanisms most likely to produce good results in pursuing goals may differ sharply from those that are effective in promoting compliance with rules and regulations. The success of goal-setting, for example, generally requires building a coalition dedicated to making progress, whereas eliciting compliance from individual subjects is essential to the success of rule-making. The effectiveness of goal-setting as a governance strategy may vary as well to the extent that the strategy is used as a stand-alone tool for solving governance problems, as a supplement to a rule-based system, or as a way of identifying targets articulated in a rule-based system.

Opportunity costs/benefits—A focus on goal-setting can deflect attention and divert resources from efforts to build effective governance systems via the creation and implementation of rule-based regimes dealing with issues like climate change or the loss of biological diversity. In cases where resources are limited, it may be important to confront choices regarding the menu of goal-setting and rule-making options. On the other hand, there may be opportunities to achieve synergy either by using these tools sequentially or employing them together.

Complacency—More generally, goal-setting can breed a kind of complacency rooted in a sense that once appealing goals are established, there is no need to invest time and energy in more demanding forms of problem-solving. This creates a danger that those who are unable or unwilling to address governance needs through regulatory measures will use goal-setting as a diversion to deflect public attention from their unwillingness to take governance problems seriously.

Determinants of the Effectiveness of Goal-Setting

The challenge of evaluating the effectiveness of goal-setting has much in common with the parallel challenge in the case of rule-making.[2] To begin with, the familiar distinction among outputs, outcomes, and impacts applies just as well to the evaluation of success regarding goal-setting as it does to the effectiveness of rule-making. Outputs relating to goal-setting include the articulation of targets and indicators associated with specific goals as well as the establishment of organizational arrangements to oversee the effort to attain goals. Outcomes refer to behavioral adjustments on the part of both states and nonstate actors intended to promote progress toward goal attainment. Impacts, then, involve progress toward fulfilling the goals themselves. As in the case of rule-making, the causal chain becomes longer, more complex, and harder to pin down as we move from outputs to impacts. It is relatively easy to establish a causal connection between the articulation of goals and the establishment of organizational arrangements designed to promote their attainment. It is another matter to demonstrate such a connection between goal-setting and actual progress toward fulfilling the relevant goals. Take the case of MDG 1 focusing on poverty eradication as an example in this regard. As a number of analysts have observed, there has been significant progress in eradicating extreme poverty in the period since the adoption of the Millennium Declaration in 2000 (Sachs 2015; United Nations 2015). But can we show an unambiguous causal relationship between the adoption and implementation of the MDGs and this development? At a minimum, this is a situation featuring complex causality. The MDGs may have played a role of some importance in reducing extreme poverty, but so have a variety of other factors involving economic growth, social change, and public policy in developing countries like China.

That said, there is no reason to expect goal-setting to perform equally well (or poorly) in addressing needs for governance under all conditions. What may work in a close-knit community whose members are dedicated to the good of the whole, for instance, may fall flat as a steering mechanism in a group whose members are fiercely independent actors who prioritize efforts on the part of individual members to make their own way under most conditions and who focus on relative gains in assessing the results of their efforts. This suggests the value of launching an inquiry into the determinants of success with regard to the use of goal-setting as a governance strategy. Here, I set forth some preliminary observations about four important sets of conditions relating to: (i) the nature of the problem, (ii) the character of the actors, (iii) the principal features of the setting, and (iv) the mobilization of public support in specific cases.

Nature of the Problem

The problems that generate needs for governance are not all alike; some are likely to be more amenable to goal-setting strategies than others. In the case of lumpy collective goods, for instance, those seeking to supply governance must overcome not only the familiar challenge of the free-rider problem but also complications arising from the fact that none of the good can be supplied until a specific (possibly high) threshold is reached. Where collective goods are continuous, by contrast, early efforts allowing for initial supplies of the goods may encourage members of the group to make additional contributions to obtain more. In cases where the problem is spatially defined (e.g., protecting biodiversity in mega-diverse countries), there are often complex issues regarding the extent to which outsiders can or should contribute and the modalities to be used in channeling the contributions they do make. In cases where the problem is finite (e.g., eradicating a disease like malaria), fulfilling a goal may solve the problem once and for all. Where the problem is continuous (e.g., controlling emissions of greenhouse gases), on the other hand, goal-setting may prove less effective as a governance strategy, since it may be impossible to make clear-cut assessments of progress toward goal attainment. Beyond this lie differences in the scale and scope of the problem. Achieving MSY in a fishery that is spatially limited and lightly used by a few easily identifiable fishers is one thing; pursuing the same goal in a far-flung fishery heavily used by many fishers who engage in illegal, unregulated, and unreported (IUU) activities is another. This discussion introduces the now familiar concern with the problem of fit (Young, King, and Schroeder 2008). In a word, success in goal attainment will depend on the extent to which the goals selected and the procedures introduced to pursue them are compatible with the defining features of the problem they seek to solve.

Character of the Actors

Achieving success is likely to be determined in part by the extent to which the behavior of the actors is rooted in the logic of consequences versus the logic of appropriateness (March and Olsen 1998). Where the logic of consequences prevails, success will require an appeal to actors on the basis of calculations of benefits and costs. When the logic of appropriateness takes hold, by contrast, tying goal-setting to norms and principles may prove effective. More generally, goal-setting may work well in situations where the objective can be integrated into a coherent social narrative, so that it becomes part of how actors perceive their identity and organize their thinking about governance. In cases where making contributions to the common good is culturally prescribed, for example, it may be comparatively easy to attain goals that involve the supply of public or collective goods. At the international level, there is the added challenge of dealing with actors that are themselves collective entities. Not only does this introduce the familiar issue of two-level games in thinking about the effectiveness of goal-setting; it also makes it important to think about a host of issues arising from the facts that living up to international goals may prove highly contentious in the domestic politics of individual states and that commitments to such goals may wax and wane over time as governments come and go at the domestic level. A particularly important challenge arises when a new administration takes office seeking to differentiate itself from its predecessor and looking for ways to free up resources to provide the wherewithal to launch new policy initiatives. Forces of this sort undoubtedly play a role in the common failure of developed countries to fulfill the pledges they make regarding financial assistance to developing countries needed to attain goals like protecting the Earth's climate system or reducing the loss of biological diversity.

Features of the Setting

Goal attainment is commonly affected by features of the prevailing social setting, such as the number of actors involved, the extent to which the actors are linked together through ties of common interests or cultural affinity, the degree to which affluence eases the burden of making contributions toward the attainment of common goals, and the prospect of technological innovations that are likely to prove helpful in problem-solving. Settings involving large numbers of actors for whom attaining common goals would necessitate major sacrifices regarding the pursuit of individual goals, for example, are not conducive to the use of goal-setting as a governance strategy. It is too easy for individual actors to conclude that whatever they do will not make a significant difference in such settings. Technological innovations, on the other hand, can alleviate problems that loomed large prior to their development (e.g., dealing with congestion in uses of the geomagnetic spectrum, tracking compliance on the part of individual states or private actors). In such cases, goal attainment may prove much easier than expected at the time of the articulation of the goals. Beyond this lie issues of community, culture, and shared history in the pursuit of common goals. Where trust based on a long history of working together to address common concerns is present, for example, goal-setting may become a routine method of problem-solving. Where long-standing antipathies breed distrust and efforts to work together are apt to generate misunderstandings, by contrast, goal-setting rarely proves effective as a means of addressing needs for governance.

A particularly relevant consideration in some settings centers on the influence of ideology or dominant social narratives. There is some tendency to associate goal-setting with centrally planned or socialist systems in which the state is not only able to establish goals or targets but also to take effective steps to allocate resources or factors of production needed to attain goals. Rule-making, on the other hand, is often associated with (neo)liberal systems in which the state can establish the rules of the game but is otherwise expected to minimize intrusions into the affairs of private actors and to rely on regulations applicable to all (or all members of a certain class of actors) when it becomes necessary to take action to promote the public good. It is important not to exaggerate the significance of this distinction. Consider the role of goal-setting in the United States in cases like the Manhattan Project during WWII or the Apollo Project during the 1960s, as well as the role of environmental regulations in centrally-planned systems like China. Nevertheless, it is worth thinking carefully about the prospect that the reactions of key players to measures featuring goal-setting will be influenced by ideological considerations.

Mobilization of Support/Leadership

Those who make use of goal-setting as a governance strategy in specific situations may be more or less successful in mobilizing and sustaining the support needed to attain their goals. In part, this is a matter of building coalitions of the willing, highlighting the advantages of joining coalitions of supporters, and promising various sorts of rewards for those who not only do their part in fulfilling goals but also encourage others to do so. Partly, it is a matter that involves a kind of championship or leadership on the part of influential individuals who are able to present goals in an appealing manner and to inspire those working to attain the goals with a sense of mission regarding the importance of their roles in the process. In either case, however, the point is that it is not sufficient simply to articulate goals and expect the members of society to make a concerted effort to fulfill them. Goal-setting as a governance strategy requires a concentrated effort to energize existing supporters and recruit new supporters to pursue the common objective, sometimes over a relatively long period of time.

Goal-Setting in the Anthropocene

How is goal-setting as a governance strategy likely to fare as we move deeper into the Anthropocene? One immediate concern is that the needs for governance arising in this setting may trigger a growing tension between the priorities of those who focus on systemic issues like climate change or problems associated with planetary boundaries more generally and those who address more familiar, proximate issues like eradicating poverty or wiping out a particular disease. There is little overlap between the communities addressing these different types of issues. A typical response is to treat these diverse concerns as discrete issue areas and to develop concrete goals in each area with little attention to coordinating the two processes (Young and Steffen 2009). But this response is not likely to produce the desired results in the Anthropocene not only because staying within a "safe operating space for humanity" is a necessary condition for success in dealing with proximate issues, but also because resource constraints ensure that pursuing specific goals will generate opportunity costs regarding the capacity to address others effectively (Rockström et al. 2009; Rockström and Klum 2015; Steffen et al. 2015). In effect, this is likely to make the priority-setting needed to govern effectively through goal-setting more difficult. Initiatives that can be presented credibly as win-win options with regard to the two types of issues will be particularly appealing in this context.

Using goal-setting as a strategy for addressing systemic concerns like climate change also raises questions about the proper way to assess performance. This is not just a problem of measurement or data collection, although it is undoubtedly easier to track changes in the concentration of carbon dioxide in the Earth's atmosphere than to track changes in the number of people living on less than a dollar a day. The problem lies at a deeper level. As the case of climate change makes clear, there is great uncertainty in identifying tractable goals regarding many systemic concerns. Treating the goal of avoiding "dangerous anthropogenic interference with the climate system" as a matter of limiting temperature increases at the Earth's surface to less than 2°C or even to 1.5°C is ultimately arbitrary (UNFCCC 1992, Art. 2; Paris Agreement 2015). What is more, we do not know with any certainty what this means in terms of limits on concentrations of greenhouse gases in the Earth's atmosphere. Nor is this case unusual. It follows that there is an inevitable element of subjectivity in setting such goals, so that it is not surprising that there is considerable variation between the preferences of those who adopt cautious attitudes and those who assume that caution is not called for because we will find ways to solve systemic problems like climate change as the need arises (Sabin 2014).

Goal-setting may seem preferable to rule-making when it comes to coping with uncertainty and enhancing adaptiveness in the face of nonlinear and sometimes abrupt and surprising changes. Rules are intended to remain in place indefinitely, and they often give rise to rigid bureaucracies dedicated to their implementation without regard to changing circumstances. Goals, by contrast, are apt to be temporally limited. The usual idea is to fulfill initial goals within a more or less fixed period of time and then to go on to set new goals during subsequent time periods. Yet this seeming advantage may be more apparent than real. Those who have invested substantial time and energy in the pursuit of a particular goal may be committed to a certain course of action in the sense that it is costly in both material and psychological terms to jump from one track to another as circumstances change. This is one reason why some analysts adopt a more incremental approach, arguing that it is better to focus on the development of processes that produce sensible or appealing results at each step along the way, rather than on the fulfillment of goals that may require a commitment to a certain outcome regardless of the implications of changing circumstances during the interval between setting a goal and fulfilling it. The relative merits of these opposing perspectives are by no means clear. But these observations do suggest that it is likely to be hazardous in the Anthropocene to place too much emphasis on the pursuit of long-term goals that yield few if any payoffs until the goal itself is reached.

Enhancing the Effectiveness of Goal-Setting

What options are available for those seeking to enhance the effectiveness of goal-setting as a governance strategy at the international level? The essential challenge here is to influence the behavior of relevant actors by sharpening their understanding of needs for governance, strengthening the commitments they make to pursuing key goals, and providing them with powerful incentives to live up to their commitments.

In some cases, such efforts may feature incentives that can be couched in terms of a utilitarian calculus of benefits and costs. Think of some of the suggestions of analysts like Schelling regarding committal tactics in situations where incentives to defect may be strong and credibility problems are therefore prominent (Schelling 1960). An actor may agree to incur costs in the event of failure to live up to a pledge, for instance, as a means of making its commitment both real to itself and credible to others.[3] There are also cases involving incentives that seem more credible to some than to others. Religious organizations, for example, regularly persuade the faithful to live up to demanding commitments (e.g., tithing) by assuring them that doing so will entitle them to enjoy rewards in heaven. While this mechanism will not appeal to those who do not believe in an afterlife, it is undeniably effective in guiding the behavior of believers.

On the other hand, there are cases in which techniques for enhancing the effectiveness of goal-setting rely on mechanisms that are hard to frame in terms of benefit/cost calculations. This is especially true in situations involving collective-action problems relating to the supply of public goods where there are the familiar incentives to defect or to become a free rider (Olson 1965). Here, it may make sense to rely on mechanisms that involve factors like honor, moral obligation, face saving, a sense of group solidarity, or even the force of habit (Hart 1961).

What measures are likely to increase the effectiveness of goal-setting in the anarchic setting of international society where many of the conventional procedures for enhancing the effectiveness of goal-setting are of limited value? Here are some observations about a range of (by no means mutually exclusive) procedures that may prove helpful in this setting.

Publicize the goals in dramatic forms—A goal like halving "the proportion of people whose income is less than US $1 a day" by 2015 (MDG 1) lends itself to formulation as a sound bite that is easy to understand and to present as a challenge to one and all (UN General Assembly 2000). It is also easy to devise a visual barometer that allows everyone to follow progress (or lack of progress) toward fulfilling such a goal on a continuous basis and to determine whether benchmarks along the way are being met. Compare this case with the goal of avoiding "dangerous anthropogenic interference with the climate system" which involves an imprecise target and no obvious measurement procedure. As recent experience makes clear, it is even possible for some actors to gain traction by denying that the problem of climate change is a consequence of human actions.

Memorialize the goals in a high-profile document or declaration— Although UN General Assembly resolutions are not legally binding on member states, they can become high profile documents that serve to increase the visibility of goals, to provide key goals with a sense of legitimacy, and to draw attention to the extent to which actors contribute toward attaining them. The UNGA's Millennium Declaration launching the MDGs in 2000 is a prominent example (UN General Assembly 2000). The General Assembly has adopted a similar procedure regarding the SDGs, launching them officially in a September 2015 UNGA resolution (UN General Assembly 2015).

Formalize commitments—It may help to formalize commitments even when they are not legally binding. Fundraisers are familiar with this mechanism. Those who agree to contribute US $100 a month to some worthy cause are likely to get into the habit of doing so on a regular basis, even when they are not legally obligated to make good on these commitments.[4] An important principle of fundraising is that those who have given already are the most likely candidates to provide additional contributions. They may even routinize their contributions by authorizing automatic charges to bank accounts or credit cards. In such cases, it will take a conscious act to depart from the path leading toward goal attainment; path dependence then works in favor of success.

Make formal pledges so that non-performance will cause embarrassment or loss of face—The idea here is that actors are likely to avoid the embarrassment of reneging on pledges, even when they are under no formal obligation to make good on them. The pledges that countries were called upon to make under the terms of the 2009 Copenhagen Accord on climate change provide a prominent example of this mechanism (Copenhagen Accord 2009). Many have criticized the Copenhagen Accord on the grounds that these pledges are essentially voluntary in character. Yet it is interesting to observe the extent to which leaders feel some sense of obligation about at least attempting to fulfill their pledges. The prominence of what are now known as INDCs, which have become a central focus of efforts to reduce emissions of greenhouse gases, reflects a rising interest in goal-setting as a governance strategy; the results will constitute a fascinating test case of the effectiveness of goal-setting in international society.

Launch a social movement dedicated to attaining the goal—While protecting the climate system from dangerous anthropogenic interference may be an amorphous goal, the goal of the movement calling itself 350.org is clear and understandable to all; it is easy to monitor movement either toward or away from attaining a goal stated in terms of concentrations of carbon dioxide in the Earth's atmosphere. Equally important, the goal of limiting the concentration of CO_2 in the atmosphere to 350 ppm has become the rallying cry for a social movement galvanizing the actions of people around the world (McKibben 2013). Whatever their merits from the perspective of rational choice, the actions of social movements can trigger profound changes in societies (e.g., the abolition of slavery, the granting of universal suffrage), especially when the goals of such movements are easy to understand and progress toward goal attainment is easy to monitor (Klein 2014).

Make the goals legally binding—Giving goals the force of law may increase the willingness of some actors to live up to their commitments or pledges on the grounds that legal obligations have a normative pull of their own. The idea here is that making actions legally mandatory can influence behavior even when there are no formal sanctions or the penalties for non-compliance are modest. The goal of avoiding "dangerous anthropogenic interference with the climate system," articulated in Art. 2 of the UNFCCC, is a prominent example. Yet it is important to note as well that the behavioral significance of a sense of legal obligation is influenced by broader cultural perspectives that are likely to vary from place to place and across time.

Establish well-defined benchmarks for assessing progress—In addition to developing indicators for measuring progress, it often helps to establish an explicit timetable for making progress toward reaching goals and to define benchmarks to be used in assessing whether efforts to attain goals are on track. Such measures have the effect of subdividing the overall goal into more manageable chunks and establishing checkpoints that facilitate efforts to assess progress and that allow for mid-course corrections if necessary. Especially in cases where achieving an overall goal is apt to be a lengthy process, some actors will find it helpful to have definite benchmarks that can serve as interim targets in addition to the final goal.

Tie other goals/receipt of rewards to the fulfillment of the goals—Yet another procedure is to make fulfillment of initial goals a requirement for moving on to the pursuit of higher order and highly valued objectives. This is a common procedure at the individual level, where promotion or advancement to a higher status or rank is commonly predicated on the fulfillment of more or less specific goals treated as qualifiers. Think of familiar situations in which passing specific courses is a prerequisite for enrollment in higher-level courses or acquiring some sort of certification is needed to apply for a license to practice a profession. At the international level, an interesting illustration of this mechanism is the requirement of meeting various intermediate goals as a precondition for admission to membership in the European Union.

The MDGs and the SDGs as Examples of Goal-Setting

To illustrate the applicability of the preceding analysis to real-world situations, I turn now to a more focused discussion of the MDGs and the SDGs. At the outset, it is useful to draw a distinction between the MDGs and the SDGs as goal-setting exercises. The idea of the MDGs arose during the 1990s as a means of responding to the concerns of developing countries during a period featuring an emphasis on overarching or systemic issues like climate change and biodiversity of particular interest to advanced industrial countries. This accounts for the emphasis in the MDGs on concrete issues, such as eradicating poverty, improving sanitation, and providing primary education, of special concern to developing countries. In effect, the MDGs constituted one side of a global political bargain (Young and Steffen 2009).

The effort to develop and implement a set of sustainable development goals represents a different and more ambitious undertaking. As the efforts of the Open Working Group created to follow up on the Rio+20 mandate suggest, many of those engaged in this work sought to emphasize continuity between the MDGs and the SDGs (Open Working Group on Sustainable Development Goals 2014). Critical problems involving poverty, food security, basic human health, and so forth certainly have not gone away. Nevertheless, the fundamental challenge in formulating the SDGs was to find a way to balance these ongoing concerns with growing systemic challenges in order to make progress toward integrating the social, economic, and environmental elements of sustainable development under conditions in which the impacts of human actions have become significant at the planetary level (Sachs 2015, Ch. 14). This presents a problem in part because there is no consensus regarding the meaning of sustainable development itself at the operational level, much less regarding the implications of the onset of the Anthropocene for the pursuit of sustainability. But what is clear already is that formulating and implementing SDGs requires a global deal acceptable to both the developing world and the advanced industrial world (Stern 2009). The politics of the situation may eventuate in the production of a set of goals whose terms turn out to be too vague to provide useful guidance to policy-makers. Nevertheless, this is not a valid reason to give up on making a concerted effort to meet this challenge.[5]

The Future We Want, the outcome document from Rio+20 states that the "... SDGs should be action-oriented, concise, and easy to communicate, limited in number, aspirational, global in nature, and universally applicable to all countries while taking into account different national realities, capacities, and levels of development and respecting national policies and priorities" (United Nations 2012, para. 247). What does my analysis of goal-setting as a governance strategy have to say about fulfilling these requirements?

To lend substance to this discussion, I turn to the recommendations set forth in two prominent reports—the *Report of the High-Level Panel of Eminent Persons on the Post-2015 Development Agenda* (hereafter the High-Level Panel Report) and the *Report of the Leadership Council of the Sustainable Development Solutions Network* (hereafter the SDSN Report)—as sources of examples (High-Level Panel of Eminent Persons 2013; Leadership Council for the Sustainable Development Solutions Network 2013). Of course, these are not the only relevant examples of the thinking that went into the formulation of the SDGs. But they were high profile contributions to the public debate regarding the formulation of the SDGs, and they are helpful for purposes of illustrating the concerns addressed in this section.

Minimize the Number of Distinct Goals

Those working to formulate the SDGs did not have the luxury of the leaders of charitable organizations, who can launch a fundraising campaign with the single goal of raising some specified sum of money, or the leaders of the US National Aeronautics and Space Administration, who were able to launch the Apollo Project with the single goal of landing a man on the moon during the decade of the 1960s. Not only is sustainable development a multidimensional concept; but also any effort to reduce sustainable development to a set of specific goals must confront political pressures to include objectives of particular interest to a range of influential stakeholders. Nevertheless, my analysis of goal-setting as a governance strategy suggests that there is a compelling case for heeding the injunction of the Rio+20 outcome document to make goals like the SDGs "limited in number." Articulating a suite of goals that are both numerous and range across a broad array of issue areas is apt to be self-defeating because it leads to competition for attention and conflict over the allocation of scarce resources. Consider in this context the High-Level Panel Report's twelve goals ranging from ending poverty and securing sustainable energy to ensuring stable and peaceful societies, or the SDSN Report's ten goals ranging from achieving development within planetary boundaries to curbing human-induced climate change and securing ecosystem services and biodiversity. It is not hard to comprehend the processes that produced these formulations. But there is little prospect of making significant progress across suites of goals that encompass a large proportion of the overarching set of human interests and aspirations. For this reason, the fact that the UNGA resolution (UNGA A/RES/70/1) launching the SDGs formally embraces seventeen distinct goals, ranging from eradicating poverty to promoting peaceful and inclusive societies, is a source of significant concern (see table 5.1).

Strike a Balance between Aspirations and Political Feasibility
Goals that are lacking in ambition may be comparatively easy to fulfill, but they are not capable of galvanizing political will on the part of societies to mount the sort of campaign needed to solve fundamental problems. Conversely, goals that are too idealistic or visionary will strike actors as beyond the realm of what is politically feasible and fail to serve as the unifying themes needed to make real progress. This is why a goal like eradicating poverty is appealing. It is obviously ambitious, but it also seems to lie within the realm of the possible, especially given the progress that was made in addressing the problem of poverty within the framework of the MDGs.[6] As a number of observers have argued, the period from 2016 to 2030 may be the moment to finish the job when it comes to putting an end to extreme poverty. The High-Level Panel Report makes this the focus of its first proposed goal. On the other hand, fulfilling some of the goals articulated in the High-Level Panel Report and the SDSN Report would require a sea change in human affairs that is difficult to foresee during the 2016–2030 timeframe. The idea of ensuring stable and peaceful societies is a case in point. So is the goal of devising effective measures to protect ecosystem services. There is nothing wrong with goals of this sort as long-term visionary objectives. But it is hard to see how they strike a suitable balance between aspirations and political feasibility for the period from 2016 to 2030.

Table 5.1

The UN Sustainable Development Goals, 2016–2030 (UNGA A/RES/70/1)

Sustainable Development Goals
Goal 1. End poverty in all its forms everywhere
Goal 2. End hunger, achieve food security and improved nutrition, and promote sustainable agriculture
Goal 3. Ensure healthy lives and promote well-being for all at all ages
Goal 4. Ensure inclusive and equitable quality and empower all women and girls
Goal 6. Ensure availability and sustainable management of water and sanitation for all
Goal 7. Ensure access to affordable and sustainable management of water and sanitation for all
Goal 8. Promote sustained, inclusive, and sustainable economic growth, full and productive employment, and decent work for all
Goal 9. Build resilient infrastructure, promote inclusive and sustainable industrialization and foster innovation
Goal 10. Reduce inequality within and among countries
Goal 11. Make cities and human settlements inclusive, safe, resilient, and sustainable
Goal 12. Ensure sustainable consumption and production patterns
Goal 13. Take urgent action to combat climate change and its impacts*
Goal 14. Conserve and sustainably use the oceans, seas, and marine resources for sustainable development
Goal 15. Protect, restore, and promote sustainable use of terrestrial ecosystems, sustainably manage forests, combat desertification, and halt and reverse land degradation, and halt biodiversity loss
Goal 16. Promote peaceful and inclusive societies for sustainable development, provide access to justice for all, and build effective, accountable, and inclusive institutions at all levels
Goal 17. Strengthen the means of implementation and revitalize the global partnership for sustainable development

* Acknowledging that the United Nations Framework Convention on Climate Change is the primary international, intergovernmental forum for the negotiating the global response to climate change.

Devise Effective Procedures to Track Progress

Once again, the contrast with the example of the capital campaign is instructive. Such a campaign has a single goal that is inherently operational. There is no need to devise an elaborate apparatus of targets and indicators to track progress toward fulfilling such a goal. In fact, it is possible to present a single chart using an analogy to a barometer to display progress toward reaching the overall goal at any time. It is also relatively easy to establish temporal benchmarks in such cases. Naturally, things are more complex when it comes to formulating and implementing the SDGs. Still, it is important to bear in mind the Rio+20 injunction to make the goals "concise and easy to communicate." This is one reason why ending poverty is an appealing goal. So long as we provide an operational definition of poverty (e.g., living on less than US $1 or $1.25 a day), it is comparatively easy to track progress toward fulfilling this goal. But other goals, ranging from achieving equitable growth to ensuring good governance and transforming governance for sustainable development, present fundamental challenges for those seeking to track progress. Partly, this is a matter of constructing operational measures, a situation that explains the elaborate and rather cumbersome effort to construct suites of targets and indicators to go along with each of the SDGs (Shepherd et al. 2015). In part, however, it presents fundamental challenges that are more normative in nature, such as determining what we mean by equity or good governance.[7]

Make Goals Appealing to Different Motives Driving Behavior

Because behavior has a variety of sources, it makes sense to formulate the SDGs in a manner that appeals to those whose actions are rooted in a range of motives. One useful distinction in this context contrasts the logic of consequences, featuring incentives linked to utilitarian calculations of benefits and costs, and the logic of appropriateness, marked by more normative concerns and matters of principle (March and Olsen 1998). Especially in cases where the goals address collective-action problems (e.g., protecting the Earth's climate system) or the need to avoid unintended (and often unforeseen) side effects (e.g., damages to ecosystem services arising from efforts to achieve food security), it is important to find ways to proceed that encourage actors to transcend narrow conceptions of self-interest and to embrace principles (e.g., the precautionary principle, the polluter pays principle) that can motivate all parties concerned to behave in ways that promote the common good, even when such behavior may prove costly in the short run (Young 2001b; Chapter 6 *infra*). This suggests an important rationale underlying familiar goals like providing high quality education, ensuring healthy lives, and promoting food security (goals 3, 4, and 5 of the High-Level Panel Report). There is a distinct sense in which all should embrace these goals on the basis of self-interest. Even the wealthy should conclude that meeting these goals is a matter of enlightened self-interest, since doing so will contribute over time to securing a safe, vibrant, and productive society that is beneficial to all.

Join Goal-Setting and Rule-Making to Create Integrated and Effective Governance Systems

There is much to be said for finding ways to join goal-setting and rule-making to maximize the effectiveness of governance systems. Goal-setting serves an aspirational function, providing actors with vision and a guiding rationale for participating in a governance system. Rule-making, on the other hand, can provide the behavioral prescriptions (i.e., the requirements and prohibitions) required to tell actors how they need to behave to make progress toward fulfilling goals. Goals in the absence of rules are apt to degenerate into vague aspirations that everyone embraces in principle but no one knows how to fulfill in practice. Rules in the absence of goals, on the other hand, are apt to degenerate into burdensome and bureaucratic requirements that no one sees as needed to achieve overarching goals. This suggests a need to do more to connect goal-setting and rule-making as governance strategies in the global effort to pursue sustainable development. At this stage, the effort to pursue the SDGs is proceeding on a separate track with little input from those working on substantive issues like the control of disease vectors, the reduction of greenhouse gas emissions, or the protection of endangered species.[8] This is not to say that the effort to formulate and implement a set of SDGs for the period 2016–2030 was misguided. But to the extent that this process operates without strong links to efforts to address a variety of substantive issues, the prospects for achieving the buy-in required to make real progress toward fulfilling suitably ambitious goals will suffer.

A Concluding Thought

The pursuit of the SDGs is fraught with pitfalls. But the effort to adopt and implement the SDGs also provides an excellent opportunity to explore the differences between goal-setting and rule-making as distinct governance strategies and to examine both the conditions under which each is likely to prove effective and the prospects for combining them in a manner that produces synergy. In an important sense, the process now underway regarding the SDGs differs significantly from the process unfolding in the 1990s that led to the articulation of the MDGs in the 2000 Millennium Declaration. Whereas the earlier process reflected a political bargain designed to engage developing countries and persuade them to join efforts to tackle issues of global environmental change (e.g., climate change, the loss of biological diversity), the current process is about devising an approach to the full range of human-environment interactions in a world of complex systems and finding ways to track accomplishments in this realm (Young and Steffen 2009). In a real sense, we could simplify the SDGs into two basic goals: complete the work of the MDGs while respecting planetary boundaries. There is no guarantee that the SDGs, as framed in the UN General Assembly's resolution on "transforming our world," will yield useful results. In retrospect, we may come to view them as a list of goals that is too long and framed in terms that are too vague to provide practical guidance. Yet the process does offer an opportunity to chart a global course in governing complex systems during an era in which 7–9 billion human beings have achieved the capacity to dominate planetary systems.

6

Principled Governance

Introduction

Are we witnessing the emergence of a meaningful system of environmental ethics at the international level? If so, will this development offer new ways of thinking about governance that are helpful in steering complex systems toward the achievement of sustainability in the Anthropocene? To ask these questions is to launch an inquiry into the nature of principles and the roles they play in governance on a large scale. My interest in addressing these questions is not to ask whether such principles ought to guide behavior in this realm or what the content of such principles ought to be. Rather, I am interested in the extent to which ethical precepts or codes of conduct do have an impact on behavior in a social setting that many regard as antithetical to the operation of ethical principles. Assuming that they do make a difference, I seek to assess the extent to which the rise of such principles influences efforts to meet the challenge of governing for sustainability in the Anthropocene. To encapsulate the central concern of this inquiry, I endeavor to evaluate whether the idea of principled governance can augment the social capital available to those wrestling with the challenges of governing complex systems.

What is a meaningful system of ethics? In this inquiry, I treat an ethical system as an interlocking set of normatively grounded principles or a code of proper conduct applicable to the behavior of actors in a recognized field of human endeavor. Legal ethics and medical ethics, for instance, are ethical systems in this sense. Ethical systems become meaningful to the extent that the principles they establish guide the behavior of identifiable groups of actors. There is no need for all members of a group to behave in ways that conform to the highest standards at all times for an ethical system to be meaningful. The relevant question is: how much importance do individual members of the group attach to ethical principles in weighing the pros and cons of the choices they are called upon to make under a variety of circumstances? Approached in this way, the meaningfulness of an ethical system is a variable whose value may range from low to high.

The principles of international environmental ethics have yet to be spelled out clearly, refined through repeated applications to specific situations, and codified in a single authoritative text. There is no universal declaration or charter of environmental ethics that all or virtually all actors in international society (including nonstate actors as well as states) acknowledge as legitimate and authoritative. Nonetheless, it is reasonable to focus in this assessment on a small number of texts that most observers treat as particularly salient and influential expressions of environmental principles intended to guide the behavior of actors at the international level. In this analysis, I pay particular attention to the ethical principles set forth in five key documents: the Stockholm Declaration on the Human Environment (1972), the World Charter for Nature (1982), the Rio Declaration on Environment and Development (1992), the Johannesburg Declaration on Sustainable Development (2002), and the outcome document from the UN Conference on Sustainable Development known as *The Future We Want* (2012) (Stockholm Declaration 1972; World Charter for Nature 1982; Rio Declaration 1992; Johannesburg Declaration 2002; United Nations 2012). A sixth document, the Earth Charter, is worthy of some consideration in this connection as well, though it is a product of civil society rather than an expression of intergovernmental agreement.[1]

In seeking to answer the questions posed at the outset, I start by extracting the major environmental principles embedded in these prominent texts. I then pose and seek to answer a series of interrelated questions about these principles. What is the nature and the status of the environmental principles set forth in these texts? In what ways and to what extent do they influence the behavior of the actors in international society? Do these principles in their current form add up to a coherent system of environmental ethics applicable throughout international society or, for that matter, global civil society? Has the content of this ethical system evolved or changed significantly over time? If so, what forces account for such changes? In answering these questions, I endeavor to shed light on the role of principled governance in efforts to meet the challenge of governing for sustainability in the Anthropocene.

The Emergence of International Environmental Principles

The Stockholm Declaration adopted at the close of the UN Conference on the Human Environment (UNCHE) in 1972 sets forth twenty-six separate principles. The Rio Declaration emanating from the UN Conference on Environment and Development (UNCED) twenty years later articulates twenty-seven. For its part, the World Charter for Nature, approved in the form of a UN General Assembly resolution in 1982, has twenty-four operative paragraphs. The Johannesburg Declaration, adopted at the close of the World Summit on Sustainable Development in 2002, reaffirms the Stockholm and Rio principles and directs attention specifically to the implementation of these normative standards. *The Future We Want,* the outcome document from the 2012 UN Conference on Sustainable Development (popularly known as Rio+20), reaffirms these principles again and pushes the discourse firmly toward the ideal of sustainable development. It is not possible here to canvass or even to summarize the full range of the emerging collection of normative standards intended to apply to human-environment relations throughout international society. Even so, a brief account of the contents of these key texts will serve to lend substance to the concept of ethical principles pertaining to international environmental matters and to provide a point of departure for the discussion to follow focusing on the analytic questions posed at the outset.

Some environmental principles have been stated and restated in a highly consistent manner throughout the last several decades. Both Principle 21 of the Stockholm Declaration and Principle 2 of the Rio Declaration, for instance, set forth—in nearly identical language—the proposition that "States have ... the sovereign right to exploit their own resources pursuant to their own environmental policies and the responsibility to ensure that activities within their jurisdiction or control do not cause damage to the environment of other States or of areas beyond the limits of national jurisdiction."[2] The World Charter for Nature articulates much the same normative standard in somewhat different language. Paragraph 22 of the Charter, for instance, states that "[t]aking fully into account the sovereignty of States over their natural resources, each State shall give effect to the provisions of the present Charter through its competent organs and in co-operation with other States." The Johannesburg Declaration concentrates on the steps needed to translate these principles from paper to practice. *The Future We Want* recalls the Stockholm Declaration and reaffirms "… all the principles of the Rio Declaration on Environment and Development, including, *inter alia,* the principle of common but differentiated responsibilities."

In other respects, however, the last fifty years have witnessed some significant developments in the normative discourse pertaining to proper conduct in the realm of human-environment relations as captured in these authoritative texts. At Stockholm, for example, issues relating to decolonization and racial discrimination were still on the minds of participants. Thus, Principle 1 of the 1972 Declaration asserts that "policies promoting or perpetuating apartheid, racial segregation, discrimination, colonial and other forms of oppression and foreign domination stand condemned and must be eliminated." For its part, the 1982 Charter is largely a product of rising concern throughout the world about the impacts of human activities on the natural environment. To illustrate, Paragraph 2 of the Charter asserts that "[t]he genetic viability of the earth shall not be compromised; the population levels of all life forms, wild and domesticated, must be at least sufficient for their survival, and to this end necessary habitats shall be safeguarded." By 1992, the pursuit of human welfare in a sustainable manner had become a theme of transcendent importance. Principle 1 of the Rio Declaration, for instance, states that "[h]uman beings are at the centre of concerns for sustainable development. They are entitled to a healthy and productive life in harmony with nature." Despite frustrations regarding the challenge of measuring sustainable development, the 2002 World Summit on Sustainable Development in Johannesburg was all about the importance of sustainable development. For its part, the outcome document from Rio+20 in 2012 continued to support a discourse focused on sustainable development, with an emphasis on the framing of a set of Sustainable Development Goals (SDGs) designed to succeed the Millennium Development Goals (MDGs).

Are there some central trends discernible in the evolution of this normative discourse running from the 1970s through the 2010s? One major development relates to the distinction between the concepts of environmental protection and sustainable development as platforms for normative thinking that overlap but that nevertheless suggest distinct approaches to human-environment relations. Characteristic of the earlier documents is a concern for environmental protection, framed in provisions like Paragraph 1 of the 1982 Charter, which states that "[n]ature shall be respected and its essential processes shall not be impaired." Contrast this with the perspective embedded in Principle 3 of the Rio Declaration, which asserts that "[t]he right to development must be fulfilled so as to equitably meet developmental and environmental needs of present and future generations," or Paragraph 11 of *The Future We Want*, which emphasizes the importance of a "… commitment to strengthening international cooperation to address persistent challenges related to sustainable development for all."

In many concrete situations, the discourses of environmental protection and sustainable development yield normative prescriptions that are the same or at least broadly compatible. Both provide ample justification, for instance, for acting in ways that protect biological diversity and show respect for the welfare of future generations. In the final analysis, however, sustainable development is an anthropocentric discourse. It conceives of development in terms of human welfare and includes a focus on economic and social concerns, even though it clearly recognizes the importance of maintaining the integrity of biophysical systems as a basis for enhancing human welfare over time (Sachs 2015).

Is there a way to synthesize these divergent but not necessarily discordant narratives? Although the language it employs is somewhat flowery, the Earth Charter represents one interesting response to this challenge: The preamble of the Charter states that "[t]o move forward we must recognize that … we are one human family and one Earth community with a common destiny." This leads directly to the conclusion that "… it is imperative that we, the peoples of the Earth, declare our responsibility to one another, to the greater community of life, and to future generations." On this account, the ethics of environmental protection and the ethics of sustainable development converge. The pursuit of human welfare presupposes an effective effort to protect major ecosystems and to implement a series of "… interdependent principles for a sustainable way of life." The widespread appeal of the more recent discussion of heeding planetary boundaries and creating "a safe operating space for humanity" reinforces the idea that protecting the environment is a necessary condition for the achievement of sustainable development (Rockström et al. 2009; Rockström and Klum 2015; Steffen et al. 2015).

The Core of International Environmental Ethics

The issue of evolution in the normative premises underlying the discourses pertaining to human-environment relations will reemerge in a later section of this chapter. But first it is important to identify with as much precision as possible a core set of ethical guidelines or standards of proper conduct in this field that have achieved widespread acceptance at the international level or are well on their way to attaining this status. Despite the absence of a single canonical statement of these principles, most observers would concur in including seven distinct principles in the core.

The Polluter Pays Principle

Perhaps the oldest and most widely shared principle of international environmental ethics centers on an acknowledgment that those whose actions cause harm to the welfare of others are responsible for the consequences of their actions. As phrased already in Principle 21 of the Stockholm Declaration, the emphasis is on responsibility for damage to the environment of "other States or of areas beyond the limits of national jurisdiction." The basic idea of this principle is clear enough. It features an extension to international society of what is commonly known as the nuisance doctrine in municipal settings and introduces, at least implicitly, the concept of liability as a normative construct applicable to international affairs. That said, however, the polluter pays principle raises a host of subsidiary issues that require attention but have not been addressed fully at the international level. Although Principle 21 refers explicitly to states, it seems reasonable to conclude that the doctrine applies also to the actions of nonstate actors, such as corporations, whose actions may affect the welfare of those residing in other jurisdictions. What is less clear is whether states are ultimately responsible for the damages caused by the actions of their nationals, including nongovernmental organizations and multinational corporations incorporated within their jurisdictions. Unclear as well are the standards of liability associated with the polluter pays principle. Does the principle presuppose a standard of strict liability in the sense that actors are responsible for the consequences of their actions regardless of knowledge or intent? Does the distinction between causal responsibility and moral responsibility make a difference in this context (Müller, Höhne, and Ellermann 2009)? Is it sufficient to compensate victims after the fact, in contrast to making a concerted effort to avoid causing harm in the first place? Are there workable guidelines concerning the calculation of damages in transboundary settings? Who represents interests that lie "beyond the limits of national jurisdiction"? The polluter pays principle does not offer any simple or straightforward answers to these questions.

Nor does international practice carry us far in understanding the operational content of this ethical principle. Many commentators have viewed the Trail Smelter case, involving damages caused by a smelter operating in British Columbia to various actors located across the international boundary in the state of Washington, as an important step in this connection (Bratspies and Miller 2006). But subsequent efforts to flesh out the polluter pays principle have proven somewhat disappointing. There are a number of instances in which victims of pollution have paid or at least shared the costs of repairing damages attributable to the actions of others. In the 1976 convention dealing with discharges of sodium chloride into the Rhine River, for instance, the Netherlands as the downstream victim shouldered a larger share of the cost of cleanup than France, the state in which the source of the pollution was located (Dupont 1993; Bernauer 1996). Finland and Sweden have found it cost-effective to provide financial assistance to Russia and several Eastern European countries in an effort to reduce the volume of airborne pollutants originating in that area but affecting ecosystems located in the Nordic countries (Hiltunen 1994). Much the same can be said of a variety of instances in which advanced industrial states have sought to assist developing countries in efforts to eliminate or mitigate transboundary movements of airborne or waterborne pollutants. None of this justifies the conclusion that the polluter pays principle is invalid as an ethical standard. There can be no doubt that it is now regarded as improper or unethical to act in ways that damage the welfare of those located in other jurisdictions. But these comments do highlight the difference between the idea of the polluter pays treated as an ethical principle and the development of a system of liability rules that are sufficiently clear and operational to be used in handling claims for compensation arising in specific situations.

The Precautionary Principle and the Corollary of Reverse Onus

The basic idea underlying the precautionary principle is straightforward. In dealing with threats to large atmospheric, marine, and terrestrial ecosystems, it is often impossible to pin down causal connections in a clear and generally accepted fashion. There is substantial evidence to support the propositions that overharvesting is a major cause of stock depletions in a variety of fisheries, that the long-range transport of sulfur dioxide and other airborne particulates damages lakes and forests in the form of acid precipitation, and that the emission of greenhouse gases (GHGs) into the atmosphere leads to potentially disruptive changes in the Earth's climate system. Yet definitive proof is often hard to come by, especially in time to support decisions about steps needed to avoid irreparable damages. In situations of this sort, the precautionary principle asserts that definitive proof regarding the relevant causal links is not required to trigger an obligation to take steps needed to prevent serious disruption of important ecosystems. The signatories to the 1987 Montreal Protocol, for instance, committed themselves to phasing out CFCs and related ozone-depleting substances before the emergence of a solid scientific consensus regarding the causal mechanism involved (Litfin 1994; Parson 2003). More recent efforts to agree on targets and timetables regarding reductions in GHG emissions constitute an even more dramatic example, since the uncertainties surrounding climate change today are considerably greater than those relating to the depletion of stratospheric ozone during the mid-1980s (IPCC Fifth Assessment Report 2014). In all these cases, the underlying message is the same. Ethical behavior at the international level requires policy-makers to act in a timely manner to deal with major environmental threats, even when significant uncertainties remain regarding the biophysical processes involved.

Stated in this form, the precautionary principle is now widely accepted; it shows up regularly in international agreements and even in legally binding conventions and treaties (Freestone and Hay 1996). The corollary of reverse onus, on the other hand, is a much more controversial prescription in ethical terms. In essence, this corollary asserts that in cases involving significant uncertainties regarding the environmental impacts of proposed actions, the burden of proof rests with proponents to show that their actions will not cause serious harm to human communities and important ecosystems rather than with the opponents to show that these actions will prove injurious. Proving beyond a reasonable doubt that a given action will be environmentally benign is typically difficult and often impossible. Consider the harvest of whales as a case in point. Proponents of the revised management procedure (RMP) have gone to great lengths to demonstrate that a controlled harvest of minke whales would pose no threat to exploited stocks of these animals (Friedheim 2001). Most thoughtful and informed observers find this reasoning convincing. Yet it remains difficult to overcome definitively objections raised by those who oppose the consumptive use of whales under any circumstances. In such cases, it becomes important to ask: what constitutes a reasonable doubt, and who should be authorized to resolve disagreements about matters of this sort? Given the nature of ethical systems in contrast to legal systems, this is ultimately a matter to be settled by individual actors who must live with the consequences of their actions. This may strike some as evidence of the weakness of ethical principles. Yet the intensity of the controversy surrounding the corollary of reverse onus is itself a clear indication of the perceived importance of ethical principles, even in a highly decentralized setting of the sort exemplified by international society.

The Principle of Environmental Equity

Environmental equity is, first and foremost, a matter of taking steps to ensure that the rich and powerful do not insulate themselves from environmental harm largely by displacing problems onto the poor and weak. Such concerns arise regularly within societies in such forms as debates over the siting of facilities for storing wastes and industrial facilities involving the use of hazardous chemicals or the generation of toxic wastes (Comacho 1998). Given the widening gap between the rich and the poor in many societies, concern about the environmental victimization of the poor has emerged as a focus of growing importance at the policy level (Dowie 1995). In international society, the principle of environmental equity applies to relations between and among states or whole societies. But the fundamental issues are much the same (Brown Weiss 1995). Three distinct sets of concerns regarding international environmental equity have come into focus in recent years. One has to do with opposition to the exploitation of developing countries as sites for the disposal or reprocessing of hazardous wastes. Much of the opposition to key provisions of the Basel Convention on transboundary movement of hazardous wastes, for instance, rests on ethical sentiments of this sort. A second set of concerns centers on questions about the appropriateness or propriety of investing scarce resources in combating problems like climate change when many of the world's poorest countries lack safe drinking water, adequate sanitation facilities, and even secure food supplies. The issue here involves real or perceived tradeoffs between tackling a future-oriented problem like climate change and acting promptly to address the immediate concerns articulated in the Millennium Development Goals (MDGs) and now included in revised terms in the Sustainable Development Goals (SDGs) (Young and Steffen 2009). Beyond this lies a set of questions encompassing issues that have to do with the provision of assistance to poor countries to allow them to participate effectively in global environmental regimes (Keohane and Levy 1996). Such considerations are clearly reflected, for example, in the creation in 1990 of the Montreal Protocol Multilateral Fund as a means of providing financial assistance to developing countries in return for their active participation in the regime created to protect the stratospheric ozone layer (Benedick 1998).

A dramatic current issue involving environmental equity centers on the effort to find ways to reduce or curb GHG emissions. Most developing countries have taken the view that the problem of climate change is a consequence of the actions of advanced industrial countries, so that it is unfair to expect them to join any effort to protect the Earth's climate system unless and until the wealthy countries take effective steps to come to terms with this problem. For their part, the wealthy countries, noting that large developing countries like China and India have become leading sources of GHG emissions, are anxious to find ways to persuade the developing countries to join in efforts to control emissions sooner rather than later.[3] In part, this is a simple matter of bargaining. Perceiving that they have real leverage with regard to the issue of climate change, countries like China and India are determined to drive a hard bargain in their dealings with the advanced industrial countries on this issue. If anything, their determination has been strengthened by the relative failure of their efforts to gain support for a "new international economic order" (NIEO) during the 1970s (Hart 1983). But it would be a mistake to overlook the ethical dimension of this issue. There is clearly a sense in which the views of those representing developing countries are shaped both by a feeling that it is improper for them to be asked to pitch in to solve a problem arising from the behavior of others and by a conviction that it is unfair to ask them to forego the use of development strategies that the industrialized countries have used successfully (Miller 1995; Mattoo and Subramanian 2012). Unless leading developed countries like the United States, Japan, and Germany acknowledge the ethical basis of these views, progress toward forming the global coalition needed to deal effectively with climate change will be limited.

The Principle of Common but Differentiated Responsibilities

The principle of common but differentiated responsibilities is an outgrowth of the principle of environmental equity. The essential idea underlying this principle involves joining a general acceptance of the proposition that we are all in the same boat with respect to many large-scale environmental problems, on the one hand, with an acknowledgment that the circumstances of individual countries differ markedly, on the other. There are at least two types of circumstantial differences that are worthy of consideration in this connection. One type involves variations among states regarding actions that cause major environmental problems. The problem of climate change discussed in the preceding paragraph is a case in point. It would be absurd not to recognize the difference between the United States and a country like India with regard to responsibility for increases in concentrations of GHGs in the Earth's atmosphere. The other type of difference centers on various measures of ability to pay or capacity to contribute to solving major environmental problems. It is obvious that poor countries, which are preoccupied with domestic problems like providing for the basic needs of their own citizens, are not in a position to make large contributions to efforts to solve transboundary or global environmental problems. The central thrust of the principle of common but differentiated responsibilities in situations of this kind is to couple an acknowledgment that everyone bears some responsibility for coping with large-scale environmental problems with a recognition of the fact that some members of international society are much better situated than others to provide the resources needed to address these problems. The principle combines a universal ethical standard with a pragmatic acceptance of marked differences in the material circumstances of individual members of international society.

In practice, many difficulties arise in efforts to apply this principle to specific situations. One way to operationalize the idea of differentiated responsibilities is to develop a sliding scale of contributions along the lines of the scale used to calculate contributions to the general fund of the United Nations. Yet experience with devising and periodically reforming such scales is far from reassuring from an ethical point of view. What is more, there are many cases in which the essential issue involves finding ways to help others to meet their obligations rather than agreeing on a specific scale of contributions. Some efforts to come to terms with such matters have worked reasonably well. The role of the Multilateral Fund in the case of ozone is a positive example. Some of the activities of the Global Environment Facility (GEF) with regard to the protection of biological diversity and the control of GHG emissions are worthy of note in this connection as well (Sand 1995). But there are major problems in living up to this element of the principle of common but differentiated responsibilities (Keohane and Levy 1996). At the UNCED conference in 1992, for instance, there was a recognition of the need for the advanced industrial countries to provide "new and additional" assistance to developing countries ready and willing to make a serious commitment to coming to terms with issues such as the protection of biological diversity. But the actual performance of leading countries like the United States in these areas during the years that have elapsed since the close of UNCED can only be described as disappointing (Shabecoff 1996). Similar comments apply to more recent promises to create a fund to help developing countries adapt to the impacts of climate change. This does not call into question the validity of the principle of common but differentiated responsibilities. But it does underline the magnitude of the gap between the ideal and the actual in this case.

The Principle of Obligation to Future Generations
There is nothing new about the idea that those alive today have an ethical or moral obligation to pass on biophysical systems to the members of future generations that are as productive and resilient as those they inherit from their predecessors (Sikora and Barry 1978). The essential idea here is that, contrary to what some legal systems suggest about the

ownership of land and natural resources, the members of each generation are temporary residents of the Earth who are entitled to use these resources to enhance their own welfare but only on the condition that their activities do not impair the ability of members of future generations to use the same resources to pursue their own welfare.[4] Many cultures have developed versions of this principle. The indigenous peoples of North America, for instance, often emphasize the need to act in a way that is sensitive to the welfare of the seventh generation beyond the present. Others have suggested an approach in which each generation is seen as owing a debt to the preceding generation that it can only discharge by making a concerted effort to leave ecosystems in good condition for the members of the next generation (Rothenberg 1993). However we choose to frame this principle, its ethical implications are clear. The fact that current users of natural resources and environmental services will disappear from the scene in the future makes it exceedingly difficult to devise any ordinary system of sanctions designed to constrain the behavior of these users in the interests of securing or even enhancing the welfare of future users. Yet the limitations of utilitarian considerations of this sort do not weaken the influence of the obligation to members of future generations treated as an ethical precept.

What does require comment here is the proposition that the principle of obligation to future generations has become an ethical standard operative in international society. In essence, this development is a simple corollary of the growing capacity of humans to cause lasting damage to ecosystems on a very large scale. Consider the ethical issues surrounding the use of nuclear energy in this connection. As the 1986 Chernobyl catastrophe and the 2011 Fukushima disaster have made abundantly clear, nuclear accidents can produce dramatic transboundary impacts whose consequences will affect the welfare of large numbers of people over a period of hundreds or even thousands of years. In the face of dangers of this magnitude, it is understandable that even policy-makers who think of themselves as realists and who make decisions grounded in the realities of power politics find themselves thinking hard about the ethical dimensions of their choices and backing away from some actions that seem attractive in terms of their short-run political consequences. Similar remarks are in order about human actions that are likely to drive numerous species to extinction and, in the process, alter forever the composition of the Earth's major ecosystems (Kolbert 2014). In an important sense, the emergence of an obligation to future generations formulated as an international environmental principle is a reflection of the scale of many human actions undertaken in today's world and the advent of what observers now describe as human-dominated ecosystems (Vitousek et al. 1997; Rockström et al. 1997). Where it once was sufficient to apply this principle to the actions of individuals or local communities, it now seems appropriate to bring the same principle to bear in assessing the actions of states and other large and powerful actors such as multinational corporations.

The Principle of Stewardship

The idea of stewardship points to the unique role that humans play as actors in biophysical systems on a planetary scale. It is pointless to ask members of other species to behave in ways that demonstrate a concern for the maintenance of large ecosystems; they are not moral agents in the ordinary sense of the term. In many cases, it is possible to justify a concern for the maintenance of ecosystems in purely utilitarian terms. Those who deplete stocks of living resources today, for example, will be faced with more or less severe shortages of these resources tomorrow. Those who pollute ecosystems in the short run may well suffer from the health effects arising from pollution in the future. Important as these utilitarian considerations undoubtedly are in some cases, it seems clear that the idea of stewardship is now emerging as a more general ethical principle. The premise here is that the uniqueness of the role that humans play as moral agents carries with it an ethical responsibility to look after the welfare of physical and especially biological systems that goes well beyond simple utilitarian calculations concerning the specific benefits accruing to humans from the maintenance of healthy ecosystems (Passmore 1994). Embedded in this perspective is a clear sense that those whose superior cognitive capacity allows them to dominate the Earth must also accept a special obligation to sustain those ecosystems that are critical to the maintenance of the planet's life support systems.

The principle of stewardship should be distinguished from two other ethical perspectives that have received a good deal of attention in general discussions of environmental ethics. One is the idea that at least the higher orders of animals (e.g., whales and wolves) have a right to life, a proposition that would make it unethical for humans to kill these animals intentionally or to engage in actions whose side-effects are likely to deprive them of life.[5] This ethical position surfaces at the international level with some regularity, just as it does in various domestic settings. Yet there is little evidence to suggest that respect for the life of non-human organisms is emerging as a universally accepted principle of international environmental ethics at this stage. Nor is any such perspective implicit in the principle of stewardship, which is perfectly compatible with the harvesting of fish and game, so long as these activities are carried out in a manner that is sustainable and humane. Another proposal centers on Aldo Leopold's concept of biotic citizenship, which treats humans as members of large ecosystems or "biotic citizens" and seeks to shift attention from the pursuit of social welfare to the pursuit of biotic welfare. This perspective leads, in Leopold's well-known formulation, to the dictum that "[a] thing is right when it tends to preserve the integrity, stability, and beauty of the biotic community" (Leopold 1996, 262). Whatever the general attractions of this way of approaching environmental ethics, there is little reason to believe that such a shift from anthropocentric to biocentric perspectives is under way at the international level today. The rapid growth of the world's human population is sufficient by itself to ensure that anthropocentric perspectives on sustainability are likely to dominate the international discourse on environmental ethics for the foreseeable future. Yet none of this undermines the validity of the principle of stewardship, which enjoins humans to act as moral agents and behave in ways that do not drastically disrupt the functioning of the Earth's major ecosystems (Chapin, Kofinas, and Folke 2009; Chapin et al. 2015).

The Principle of Caring for the Earth

Stewardship is an idea that applies to human-environment relationships regardless of the circumstances in which they occur. But today it has become apparent that human actions have reached a stage where they can affect the operation of major components of the Earth's atmosphere, biosphere, and hydrosphere that are vital not only to human life but also to other life forms on the planet (Steffen et al. 2004). Partly, this is a matter of dramatic growth in human population, which has now passed the 7 billion mark and is expected to grow to as much as 10 billion before reaching equilibrium. In part, it is a result of technological developments that allow humans to make far greater demands on the Earth's resources than they did in the past (Steffen et al. 2015).

Not surprisingly, these developments have triggered the growth of new ways of thinking about human-environment relationships. We now find ourselves devoting more and more attention to the dynamics of human-dominated ecosystems and the role of anthropogenic drivers as major forces affecting the evolution of large atmospheric, marine, and terrestrial ecosystems (Stern, Young, and Drukman 1992). Such developments have led as well to the emergence of more evocative ideas, like the vision of spaceship earth, the concept of the limits to growth, and the idea of the Anthropocene (Meadows et al. 1972; Steffen et al. 2007). Clearly, there is much that we do not understand about the role of humans as the dominant players in large biophysical systems. Much remains to be done, for instance, in identifying conditions that determine whether various practices involving consumptive uses of living resources are sustainable or not. Yet it is hard to avoid the conclusion that this dramatic rise in the role of humans as dominant players in large biophysical systems is raising ethical issues that did not require serious consideration in earlier eras.

The result in ethical terms has been a striking increase in the development of normative precepts centered on the idea of caring for the Earth and its vital life support systems (Griggs et al. 2013). As the Earth Charter puts it, "[t]he resilience of the community of life and the well-being of humanity depend upon preserving a healthy biosphere with all its ecological systems, a rich variety of plants and animals, fertile soils, pure waters, and clean air." The result is a growing insistence that "[t]he protection of the Earth's vitality, diversity, and beauty is a sacred trust." For some, these ethical precepts are undergirded and strengthened by the idea that the Earth itself is a living organism that must be accorded respect and cared for as an entity that deserves compassionate treatment in its own right (Lovelock 1979). But it is not necessary to rely on this perspective—sometimes characterized as the Gaia hypothesis—to frame a set of ethical standards centered on the principle of caring for the Earth. As in a number of other cases, it is perfectly possible to formulate a utilitarian rationale for adopting the basic argument underlying the principle of caring for the Earth. Some will see this logic of consequences as a sufficient basis for engaging in appropriate conduct in this realm. But others point to the role of ethical principles, undergirded by a logic of appropriateness, as behavioral guidelines whose influence extends well beyond the realm of utilitarian calculations and which may prove more effective in shaping behavior than injunctions requiring actors to weigh the benefits and costs associated with the options confronting them (March and Olsen 1998). The premise embedded in this principle is that caring for the Earth is always the right thing to do, so there is no need to resort to elaborate computations of costs and benefits to provide a convincing rationale for the rectitude of this course of action (Rockström et al. 2009).

The Nature of International Environmental Principles

What is the nature of these ethical principles whose formulation and codification have become a focus of animated debates and vigorous negotiations on the part of sizable groups of actors at the international level? Have some of these principles (e.g., Principle 21 of the Stockholm Declaration) passed into customary international law, thereby acquiring the force of law with regard to the behavior of identifiable groups of subjects? Does the articulation of these principles in formal declarations or resolutions of the UN General Assembly signify a high level of political commitment on the part of sponsors and signatories to their implementation? Should the answers to these questions turn out to be negative, why should we pay attention to these ethical precepts? What is it that accounts for their influence in international society?

At the most general level, ethical principles are codes of social—as opposed to antisocial—conduct that are neither legally binding nor backed by ironclad political commitments but that nevertheless have demonstrable consequences for the ways in which key actors identify current issues, frame them for consideration in policy arenas, and go about evaluating ways to address them (Kingdon 1995). Principles, on this account, are not simply primitive, underdeveloped, or evolving rules that stand in need of further refinement and "upgrading" in order to move them over time toward the status of legally binding prescriptions. It follows that principled governance is not simply a substandard form of regulatory governance.

Rules are more or less well-defined behavioral prescriptions spelling out prohibitions, requirements, and permissions that members of specified groups of subjects are expected to comply with on a regular basis

(Young 1979; Ostrom 1990). We know from domestic experience that rules—at least in their initial and more general formulations—are often articulated in ways that leave a lot to be desired in providing guidance that is sufficiently explicit or precise to prove helpful in directing the actions of individual subjects in specific situations. The resultant need to make a transition from paper to practice is what gives rise to extended efforts to devise regulations to operationalize rules as well as to the elaborate and time-consuming procedures often associated with the promulgation and interpretation of regulations. Yet none of this alters the fact that rules are prescriptions calling for compliance on the part of identifiable subjects under more or less well defined circumstances and with some expectation that violators can expect to be sanctioned.

Principles, by contrast, are normative precepts treated as guides to proper conduct. Principles may or may not determine the choices actors ultimately make in specific situations. But the assumption is that they will increase the weight of normative considerations as an actor assesses the options or courses of action available in any given situation. A failure to act in a way that conforms to an ethical principle does not constitute a violation or a breach likely to be met with formal sanctions. Nevertheless, a failure to conform to the dictates of such principles or to provide a convincing rationale for not conforming is likely to generate social opprobrium; it may give rise to a reputation for unethical behavior that is harmful to an actor in ongoing interactions with others.

It would be a mistake to assume that the articulation of principles in the form of declarations or resolutions of the UN General Assembly implies that, in crafting and approving the language of these documents, their authors are committing themselves politically to a sustained effort to conform to the resultant norms. This is not to imply that the framing of environmental principles is an exercise in futility or a conspiracy on the part of diplomats to delude themselves and deceive others about the true nature of their endeavors. The fact that the authors of these documents know that they cannot count on their governments to make a sustained effort to conform to their requirements under all conditions does not constitute proof that the articulation of principles is an empty gesture. Framing and publicizing ethical principles constitutes an exercise in developing a discourse incorporating normative aspirations. The expectation is that the establishment and periodic reaffirmation of these principles will give rise to a shared outlook affecting the behavior of a wide range of actors, even if they fail to conform to the standards in some cases. Those who acknowledge the normative validity or legitimacy of ethical principles experience pressure to justify their actions in terms that emphasize the compatibility of these actions with the standards. And those who feel the need to justify their actions in this manner will find that it makes life easier to conform to the standards whenever it is not unduly difficult or costly to do so. Those who devote time and effort to the development of international principles are in the game for the long haul. They are more interested in guiding the evolution of discourses and the social practices that accompany them over time than in statistics regarding levels of compliance with specific rules.

With these distinctions in mind, we can begin to make sense out of several attributes of international principles that many observers have found puzzling or even downright confusing. For starters, it is perfectly normal to express ethical principles in generic or highly abstract terms (Sands 1995). Broadly speaking, this is standard practice in all social settings. If anything, however, the abstractness of international principles is even more striking than that of their domestic and especially their local counterparts. Consider, for instance, the principles that states have a "responsibility to ensure that activities within their jurisdiction or control do not cause damage to the environment of other States or of areas beyond the limits of national jurisdiction" and that human beings "are entitled to a healthy and productive life in harmony with nature." How much damage does it take to trigger the first of these principles? What about psychic or emotional harm in contrast to material damage? How can we separate out the causal significance of the activities of specific actors for purposes of assigning liability in complex multivariate relationships? What is the test of the ability of states to control the impacts of activities outside their jurisdiction? What measures of health and productivity are appropriate in assessing human welfare? How should we interpret the phrase "in harmony with nature"? Is it realistic to suppose that residents of advanced industrial societies do or can live in harmony with nature?

This is not just a simple matter of authoritative interpretation of the sort that is familiar in efforts to apply legal rules to specific situations. The fact is that the range of situations to which ethical principles apply is so great that there is little hope of operationalizing them through any straightforward process of interpretation. This is not necessarily a fatal flaw. Unlike rules that require compliance uniformly, principles are better understood as normative guidelines whose application is subject to situational considerations. There is no expectation that actors will comply with these principles in any simple, uniform, or routine manner. Their role is to provide direction for those seeking guideposts in terms of which to evaluate the appropriateness of alternative choices in a wide range of situations (March and Olsen 1998).

It should come as no surprise that ethical principles framed in generic terms sometimes run counter to and may actually contradict one another (Sands 1995). More or less pronounced tensions are even embedded within specific principles. Consider Principle 21 of the Stockholm Declaration/Principle 2 of the Rio Declaration in this light. This principle asserts, at one and the same time, that states are entitled to exploit their resources without interference on the part of others and that they have an obligation to do so without harming either the environment of others or the global commons. There is nothing unusual about rights that are accompanied by restrictions serving to limit the freedom of action of the rights-holders. But in this case, the assertion of the principle of sovereignty makes it hard to see how outsiders who suffer injuries can have any assurance of receiving just compensation, given that those responsible for the damages may refuse to allow outsiders access to their municipal legal systems and that international legal proceedings often require a willingness on the part of both parties to accept the jurisdiction of a relevant court (e.g., the International Court of Justice).

There are obvious possibilities as well for conflict between distinct ethical principles. The principles of environmental equity for those living today and responsibility to future generations, for example, may well clash in some situations. Given the rapid growth of the Earth's human population, it will be difficult to provide a "healthy and productive life" for those alive today, while at the same time taking steps to ensure that members of future generations can look forward to a similar quality of life. If we now add the precautionary principle to this mix, matters become even more complex. According to many who have paid attention to these issues, we should be cutting back sharply on current rates of use of natural capital in order to ensure that the Earth remains hospitable to future human inhabitants. It would be difficult to make a convincing case that the principle of environmental equity can be fulfilled simply by bringing everyone up to the level of consumption characteristic of those living in the countries of the developed world today. But the prospects for inducing residents of advanced industrial societies to cut back voluntarily on current consumption levels are dim.

There is nothing unusual about these difficulties in efforts to operationalize abstract principles and sort out tensions arising between or among them in concrete situations. Doctors face problems of this kind on a regular basis in making decisions about the use of procedures designed to prolong the life of terminally ill patients. So do lawyers who find themselves in the position of defending clients, including corporations as well as individuals, whose behavior they regard as ethically suspect and whose actions they may have reason to believe violate existing laws. It may be that these problems are particularly severe with regard to international ethics in general and international environmental ethics more specifically. International environmental problems, such as what to do about climate change or the loss of biological diversity, are extremely complex; they lend themselves to a variety of plausible interpretations from an ethical point of view. Those who make decisions on behalf of states and other international entities often find themselves confronted with severe conflicts arising from the fact that they occupy a number of distinct roles at the same time. It is tempting to opt for courses of action that promise to enhance human well-being in the short run, even when the results may be difficult to justify in terms of any reasonable standard of equity pertaining to social welfare in the long run. Still, it would be a mistake to exaggerate such differences between the operation of ethical systems in international society and the operation of similar systems in other social settings. It seems more productive at this juncture to turn to a discussion of ways to think about the effectiveness of international environmental principles.

The Effectiveness of International Environmental Principles

If international environmental principles are construed as codes of proper conduct rather than as primitive or underdeveloped rules, how should we think about their effectiveness or, in other words, the roles they play in shaping the course of events in international society and, looking to the future, in governing complex systems under conditions characteristic of the Anthropocene? We know—or think we know—what questions to ask in evaluating the effectiveness of rules (Young 1979; Raustiala and Slaughter 2002; Mitchell 2008; chapter 2 *supra)*. Have the rules been operationalized through some process of promulgating detailed regulations? What is the level of compliance with the rules on the part of members of the relevant groups of subjects? Is there some procedure available to generate authoritative interpretations of individual rules in cases where the relevant parties disagree about their operational meaning? To the extent that my account of the character of ethical principles is correct, however, it will not do to approach the issue of their effectiveness in terms of questions of this sort. We need a different approach to the assessment of the effectiveness of ethical principles. In this section, I suggest three distinct—albeit related—ways to assess the effectiveness of these principles. Although these perspectives do not lend themselves easily to the development of quantitative indicators, they nevertheless provide a basis for understanding the importance of ethical principles, even in a highly decentralized social setting of the sort prevailing at the international level (Finnemore and Sikkink 1998).

Linguistic Capital

Ethical principles influence the course of world affairs through the development of a vocabulary that actors can and do use in framing items for inclusion on the international agenda and in defining appropriate ways to address the issues embedded in them. During much of the postwar era, the dominant discourse in world affairs has featured a neo-realist emphasis on the role of states construed as unitary actors that pursue their interests through interactions guided for the most part by considerations of power in the material sense. This discourse lends itself to detailed assessments of initiatives designed to enhance the power of individual actors or to counter similar actions on the part of others (Bull 1977). But the neo-realist lexicon does not offer much scope for sophisticated efforts to expand the international agenda to include issues framed as matters of environmental equity or the obligation to care for the Earth's life support systems. In this context, the role of environmental principles is, first and foremost, to provide the resources needed to frame a new set of questions for consideration at the international level and, having succeeded in doing so, to provide a coherent ensemble of terms and concepts that policy-makers can use as they endeavor to develop persuasive answers to such questions with regard to specific situations.

No one expects the polluter pays principle to lead in the near future to an established practice in which states routinely comply with clear-cut and legally binding liability rules and regularly compensate victims located in other countries for damages arising from actions taken within their own jurisdictions. Yet the emergence of the polluter pays principle has provided a basis for talking about the environmental responsibilities of states, and more specifically the obligation to make a good faith effort to avoid causing injuries to others and to compensate those who do suffer injuries, even in a world in which the influence of the doctrine of sovereignty remains strong. Similarly, the principle of caring for the Earth provides a frame of reference for addressing issues that transcend mundane considerations of power politics. No one expects this principle to deter influential actors from engaging in various forms of power politics or to blunt the influence of geopolitics. But it does offer a way to think about questions concerning the appropriateness or legitimacy of using environmentally destructive means (e.g., detonating nuclear weapons or setting fire to oil wells) in the pursuit of ends defined in terms of national interests. The issue here has more to do with providing the vocabulary needed to frame ethical questions than with determining how actors respond to these questions in specific situations or in the short run (Litfin 1994). Powerful actors may seek to deny the relevance of ethical questions under particular circumstances. But once clear and compelling formulations of these questions make their way into the prevailing discourse, it becomes harder and harder for policy-makers to ignore them with impunity.

Non-Utilitarian Reasoning

One of the attractions of utilitarianism as an approach to decision-making is that it offers a mode of reasoning about specific choices that appears to be straightforward. Once an issue is framed and the options identified, it is a relatively straightforward matter—at least conceptually—to inquire into the benefits and costs associated with individual options, with an eye toward selecting the alternative yielding the greatest net benefit. Nonetheless, most policy-makers are well aware of the limitations of this mode of reasoning. This is partly a matter of appreciating the complications involved in making calculations of benefits and costs in situations involving multiple actors, complex interdependencies, and high levels of uncertainty. In part, however, it arises from the fact that utilitarian calculations do not offer a simple way to incorporate a variety of factors that loom large in the minds of many but that are difficult to represent in terms of conventional ways of thinking about benefits and costs (Ruggie 1998; Thaler 2015). Such factors range from the role of custom or tradition to considerations of legitimacy and propriety. Reasons of this kind can influence behavior without being introduced into policy-making processes at a conscious level. But in many cases, the problem of coming to terms with these non-utilitarian reasons to act in one way or another arises from the absence of suitable concepts in terms of which to articulate such considerations and juxtapose them to more conventional considerations expressed in terms of benefits and costs.

Under the circumstances, an important function of ethical principles is to provide structure and content to arguments about the merits of alternative courses of action that are not—and sometimes cannot be—expressed in the language of benefits and costs. The idea that the current generation should acknowledge obligations to the members of future generations has far-reaching implications for policy-making in a variety of situations, for instance, but it is not a simple matter of imputing costs to particular options that can be compared directly with (discounted) benefits accruing at some time in the future. Much the same can be said about the role of environmental equity as an ethical consideration that demands attention regardless of its impact on benefit/cost calculations. No one expects ethical reasons to trump utilitarian reasons on a regular basis and, in the process, to transform the behavior of important actors in world affairs.[6] Even so, providing a coherent language in terms of which to express non-utilitarian reasons constitutes an important step, making it harder to ignore such considerations or to dismiss them on the grounds that they do not offer behavioral guidance that is sufficiently clear to compare and contrast with preferences derived from utilitarian calculations.

Normative Benchmarks

Defining questions and providing a mode of reasoning in terms of which to address them are important contributions. But ethical principles also play a role in spelling out standards for evaluating and judging behavior after the fact. Many policy-makers are goal-oriented actors who are inclined to take the view that the ends justify the means under a wide range of circumstances. Yet it would be a mistake to assume that policy-makers are insensitive to the importance of justifying their actions in normative terms, whether or not this is a matter of assuaging their own guilty consciences or of avoiding shame in the eyes of others. Environmental principles enjoining actors to make contributions to solving common problems (e.g., controlling transboundary pollution), to pay attention to the needs of future generations, and to care for the Earth's life support systems emerge as normative benchmarks in the sense that they offer standards against which to assess actual performance. This is not a matter of passing judgment regarding compliance in any strict sense of the term, or even about conformance in a more general sense. Most ethical principles are too abstract to be used for this sort of assessment. In any case, no one expects policy-makers to treat environmental principles as prescriptive requirements or prohibitions to be complied with in any precise sense on a case-by-case basis. Nonetheless, such principles do stand as beacons that can be used to assess broad trends or the overall flow of events over time. Are we making progress in providing those residing in developing countries with safe drinking water and adequate sanitation facilities, as called for by the principle of environmental equity? Are there indications that humans are assuming increased responsibility for the maintenance of large atmospheric, marine, and terrestrial ecosystems, as mandated by the principle of stewardship? Dramatic progress regarding issues of this sort is not likely to occur over the short run. But to the extent that policy-makers anticipate that their overall records will be assessed after the fact in terms of broad normative criteria of this sort, we can expect a slow but ultimately significant shift in the flow of day-to-day behavior over time.

Do these considerations justify an affirmative response to the question posed at the outset about the development of a meaningful system of international environmental ethics? It is difficult to devise any precise way to measure the impact of environmental principles in formulating questions, providing means for expressing non-utilitarian reasons, and establishing normative benchmarks in terms of which to assess the course of events. There are good reasons to avoid jumping to overly optimistic conclusions about such matters. It would be naive to expect the requirements of the principle of environmental equity to be satisfied under the conditions of inequality prevailing in today's world, for instance, and there is every reason to be concerned about the efficacy of efforts to redirect behavior in areas highlighted by the principle of caring for the Earth.

At the same time, it would be a mistake to dismiss these international environmental principles out of hand. The importance of paying attention to the unintended consequences of actions that harm the welfare of others can no longer be denied at the international level. The proposition that uncertainty is not an adequate justification for inaction in responding to major environmental problems is now widely regarded as both legitimate and compelling. The importance of caring for the Earth is undeniable in a world in which global environmental change caused, at least in part, by human actions has become a major issue on the international agenda. There is no reason to expect the occurrence of a sharp or unambiguous turning point in this realm after which the major actors in international society will adopt or explicitly embrace a coherent set of well-defined international environmental principles. Even so, there is much to be said for the proposition that the discourse associated with these principles is steadily infiltrating the realm of world affairs and that this movement is likely to gather momentum in the coming decades.

A System of International Environmental Ethics

Do the principles discussed in the preceding sections add up to a *system* of international environmental ethics? This is not so much a question about the effectiveness of individual environmental principles as it is a query about the extent to which the collection of environmental principles I have identified forms a coherent ensemble that merits description as a code of conduct. There is no implication here that a system of environmental ethics must reflect the concerns of all those currently active in international society. In most social settings, ethical systems are particularly responsive to the concerns and preferences of societies' influential members, in contrast to the concerns and preferences of subordinate stakeholders. In the case at hand, it is notable that current discussions often place more emphasis on issues like climate change and biological diversity that seem important to the advanced industrial states than on issues like safe drinking water and adequate sanitation facilities that preoccupy many of the developing countries. An unbiased observer might react to these observations by questioning the ethics of ethical systems. This is surely a legitimate concern. But there is nothing peculiar about international society in these terms. Ethical systems are not correct or appropriate in some objective sense. They are products of social interactions, and they evolve in ways that reflect patterns of influence prevailing in the social settings in which they operate.

Existing international environmental principles have several attributes that serve to define the character of this embryonic ethical system. First and foremost, this is fundamentally a system of international ethics in the strict sense of the term, or, in other words, a set of principles applying primarily to the actions of states. In many instances, this feature of the system is articulated explicitly. Stockholm Principle 21/Rio Principle 2, for example, asserts that states have "the sovereign right to exploit their own resources ... and the responsibility" to ensure that activities within their jurisdiction do not harm outsiders. In other cases, this aspect of the principles is unambiguous, even though it is not stated in such explicit terms. The propositions that the polluter should pay and that actors should proceed with caution in dealing with large ecosystems are addressed in the first instance to national policy-makers. So also are concerns about environmental equity and caring for the Earth. This means, among other things, that states will have to address a range of questions about their responsibility for the activities of nonstate actors operating "within their jurisdiction or control." How can national governments regulate the behavior of multinational corporations whose activities generate long-range transboundary air and water pollution, and to what extent are states ultimately responsible for damages caused by the actions of their nationals? This feature of the system raises questions about the need to adjust international environmental principles to accommodate the growing role of nonstate actors and the emergence of global civil society (Wapner 1997). It is fashionable in some quarters to see the state as the problem and various nonstate actors, including both nongovernmental organizations (NGOs) and multinational corporations (MNCs), as entities that can play significant roles in solving the problem (Wapner 1996). But this seems overly optimistic. Much of the action regarding the development of international environmental ethics during the foreseeable future is likely to turn on debates about the proper roles of states and their interactions with nonstate actors in this realm.

The emerging international environmental principles typically deal with interactions between or among human actors regarding environmental matters, rather than with Leopold's idea of an ethical system focusing on "man's relation to the land and to the animals and plants which grow upon it" (Leopold 1996, 238). We are concerned, for example, with matters like the extent to which polluters are obligated to compensate those harmed by their actions, the prospect that weak states will end up as victims of the environmentally significant actions of powerful states, and the obligations of those who make decisions today to pay attention to the consequences of their actions for the welfare of future generations. All these concerns raise issues pertaining to the impact of human actions on the Earth's biophysical systems. But they are not rooted in Leopold's injunction to the effect that "[a] thing is right when it tends to preserve the integrity, stability, and beauty of the biotic community. It is wrong when it tends otherwise" (Leopold 1966, 263). The emerging system of international environmental ethics is anthropocentric in the sense that it emphasizes principles concerning the treatment of human stakeholders with regard to the environmental impacts of actions motivated by a desire to enhance individual or social welfare. This does not rule out the emergence over time of injunctions arising from the idea of biotic citizenship as a force to be reckoned with at the international level. But it seems hard to avoid the conclusion that this perspective does not constitute a key feature of the system of international environmental ethics that is emerging at the present time.

Quite apart from these attributes of the system as a whole, it is relevant to ask whether the emerging system of international environmental ethics is inclusive or comprehensive in the sense that it covers the major sets of issues relating to human-environment relations arising at the international level. Here, it is helpful to divide environmental issues into four broad categories and to comment on the applicability of emerging environmental principles in each case. This procedure yields the following taxonomy: (i) standards pertaining to internal impacts of environmental activities or *household principles*, (ii) guidelines relating to impacts of environmentally significant activities on adjacent states or *neighborhood principles*, (iii) codes of conduct relating to the use or management of shared natural resources or *principles of partnership*, and (iv) standards relating to international or global commons or *community principles*.

The international environmental principles now emerging perform unevenly in these terms. Progress is most apparent with regard to neighborhood principles and community principles. The principle of the polluter pays, for example, is based in large measure on familiar doctrines about the importance of avoiding actions likely to damage the welfare of neighbors or failing that to accept responsibility for compensating them for the damages they sustain. The injunction articulated explicitly in Stockholm Principle 21/Rio Principle 2 to ensure that activities "do not cause damage to the environment of other States" is predicated on the proposition that actors in international society are responsible for avoiding damage to their neighbors, even as they exercise their right to develop their own resources as they see fit. For their part, the precautionary principle and the principles of stewardship and caring for the Earth reflect a growing sense of the need for community principles in international society. The application of such principles is relatively straightforward with regard to the obligation to protect global systems, such as the stratospheric ozone layer and the Earth's climate system. Considerably more complex are cases, such as the protection of biological diversity, in which the development of community principles leads to injunctions involving matters normally regarded as falling within the domestic jurisdiction of individual members of international society.

As this observation suggests, meaningful household principles are particularly difficult to develop at the international level. The idea of nonintervention as an aspect of the doctrine of sovereignty remains highly influential. This is particularly true with regard to the sensitivities of small and relatively weak states that fear the prospect of environmental imperialism perpetrated by powerful states whose priorities bear little resemblance to the priorities of most developing countries (Hurrell 1992). Yet household principles are critical to a system of international environmental ethics. It is difficult to see how such a system can flourish in the absence of a well-developed code of proper conduct covering the internal environmental consequences of what the members of international society do within their own jurisdictions. For its part, the challenge of devising ethical principles to address the use of shared natural resources (e.g., rivers that cross national boundaries and migratory animals that move from one jurisdiction to another on a seasonal basis) lies in the fact that these resources often become focal points of more or less intense conflicts between or among the prospective partners (Gleick 1998). There is a clear need for ethical principles to blunt the natural tendency toward polarization in dealing with shared natural resources. From an ethical perspective, however, efforts to come to terms with this range of issues benefit neither from the obvious parallel with nuisance doctrines undergirding neighborhood principles nor from the rapid growth of intellectual capital relating to global environmental change underpinning community principles. This suggests a mixed response to the question posed at the beginning of this section. The embryonic system of international environmental ethics is state centric and rooted more in a concern for human welfare than in a concern for the well-being of the Earth's biophysical systems.

The Future of International Environmental Ethics

All ethical systems are dynamic. This is partly a matter of the rise of new issues on public agendas that call either for the development of distinct codes of conduct or the adaptation of existing standards to come to grips with emergent concerns. In part, it is a consequence of shifts in the interests and influence of major actors in specific social settings. Although the rise and fall of major actors need not lead to drastic changes in ethical systems, such systems do shift over time to reflect the viewpoints of the dominant members of specific societies. It would be wrong to infer from these observations that ethical systems evolve on a continuous and gradual basis. Such systems often develop by fits and starts in a fashion that may be described in terms of the idea of punctuated equilibrium. Existing principles can withstand a certain amount of stress without losing their normative force. But once these principles give way, a relatively rapid shift toward some alternative system of ethical standards is likely to ensue.

What can we say about the probable trajectory of international environmental ethics over the foreseeable future? And what will this mean regarding the challenge of governing complex systems? No doubt, many changes in specific principles are in store. But the fundamental debate is likely to center on differences between the ethical perspectives embedded in the discourses of environmental protection and sustainable development. To a degree, these discourses support similar or at least compatible ethical precepts. For example, they both encourage the development of the principle of caring for the Earth and the precautionary principle as ethical standards governing human-environment relations. But at bottom, the discourses of environmental protection and sustainable development rest on different premises regarding the place of human beings in the overall scheme of things. Environmental protection emphasizes biocentric perspectives that mandate a concern for the well-being of biophysical systems as a first principle, and highlight the idea of stewardship or even the concept of biotic citizenship as an overarching frame of reference in terms of which to approach specific issues involving human-environment relations. Paragraphs 1 and 2 of the World Charter for Nature, a product of the environmental debates of the 1970s and early 1980s, exemplify this perspective. Paragraph 1 lays down the foundational norm that "[n]ature shall be respected and its essential processes shall not be impaired." Paragraph 2 then seeks to flesh out this injunction by asserting that "[t]he genetic viability of the earth shall not be compromised; the population levels of all life forms, wild and domesticated, must be at least sufficient for their survival, and to this end necessary habitats shall be safeguarded."

Contrast this vision with the anthropocentric perspective of sustainable development that, at its core, mandates a concern for the well-being of biophysical systems not as an end in itself but rather as a means of securing human welfare. As the 1992 Rio Declaration puts it, "[h]uman beings are at the centre of concerns for sustainable development" (Principle 1) and "[i]n order to achieve sustainable development, environmental protection shall constitute an integral part of the development process" (Principle 4). Principles like stewardship, caring for the Earth, and a concern for future generations are important from the perspective of sustainable development. But so, too, are explicitly anthropocentric concerns like the injunctions to "cooperate in the essential task of eradicating poverty" (Principle 5) and to accord "special priority" to "the needs of developing countries, particularly the least developed and those most environmentally vulnerable" (Principle 6).

Many of those committed to the goal of developing an effective system of international environmental ethics have sought to minimize the differences between the discourses of environmental protection and sustainable development in the interests of maintaining solidarity among those actively concerned about the course of human-environment relations. Tactically speaking, this makes good sense. But given the observations set forth in the preceding paragraphs, it will come as no surprise that the occurrence of more or less severe tensions between the biocentric ethics of environmental protection and the anthropocentric ethics of sustainable development are unavoidable.

Is one of these ethical visions likely to triumph over the other during the foreseeable future? And what will this mean for those concerned about governing complex systems? There is no reason to assume that one ethical system must eventually triumph, driving out others in the process. The coexistence of ethical systems that converge on some principles but diverge on others is not uncommon. Even so, it is interesting to speculate here on the fate of environmental protection and sustainable development as alternative approaches to environmental ethics at the international level. Each approach faces serious obstacles that will impede its progress toward becoming a more effective ethical system. The discourse of environmental protection ultimately requires a shift from an anthropocentric to a biocentric mode of thought. Although a shift of this sort appeals to affluent citizens in some of the advanced industrial societies, it is difficult to see how it can become dominant in a world characterized by both a rapid growth in the population of humans and a strong desire on the part of those residing in developing countries to gain access to the material benefits associated with affluence. What is more, the growing human domination characteristic of the Anthropocene makes it imperative to concentrate on steering human actions. The idea of sustainable development, on the other hand, has proven difficult to translate from the status of a slogan that is appealing in general terms to a system of comprehensible principles dealing with more specific concerns. While it points the way toward principles highlighting concerns like environmental equity and the welfare of future generations, the influence of this perspective will remain limited unless and until its proponents succeed in constructing an ethical superstructure solidly based on the foundation provided by the basic vision of sustainable development. The effort to formulate and implement the Sustainable Development Goals speaks to this concern. But the success of this endeavor remains to be seen.

Those who approach such matters from a neo-realist point of view will naturally ask: who stands to benefit from the development of ethical systems rooted in the visions of environmental protection and sustainable development, and how influential are these stakeholders in international society? They assume that issues pertaining to ethical systems are much like other issues in the sense that the rich and powerful are likely to dominate efforts to develop and apply them. But this simple approach has serious limitations as a way of thinking about the dynamics of international environmental ethics. It is not apparent that the leading actors in international society, such as the United States, China, and the European Union, have unambiguous preferences with respect to the content of environmental principles. Those committed to the ethics of environmental protection have achieved considerable influence within the policy arenas of a number of OECD countries. Yet there is little evidence that the general public in these countries is ready to accept an ethical system rooted in biocentric principles. In any case, any serious effort to promote biocentric principles over anthropocentric principles in a system of international environmental ethics would run into profound opposition on the part of influential groups located in the developing world. Such an effort would be interpreted—with some justification—as an example of the worst sort of environmental imperialism; any serious campaign to impose biocentric principles on the developing world consequently would require a massive exercise in what Schelling and others have described as compellence in contrast to deterrence (Schelling 1966).

Beyond this, the spread of ideas in most social settings, including international society, involves processes that are difficult for powerful actors to control, even in situations featuring dramatic asymmetries in the distribution of material resources (Finnemore and Sikkink 1998). Despite the power of the United States in material terms, for example, American policy-makers have had little success in controlling the spread of ideas underlying efforts to come to grips with a range of large-scale environmental problems, such as climate change, the protection of biological diversity, and transboundary movements of hazardous wastes. This has resulted in the United States being cast repeatedly in the role of laggard, dragging its feet on initiatives designed to address major environmental problems and signing on grudgingly to agreements worked out largely by others (von Moltke 1997). It is difficult—even impossible—to solve many of today's environmental problems in the absence of a measure of cooperation on the part of the United States. But the point here is that the formulation and diffusion of ideas, including ethical principles, exhibit a dynamic that does not lend itself readily to interpretation on the basis of conventional calculations regarding the role of power in international society.

A particularly interesting aspect of this dynamic in the decentralized setting characteristic of international society concerns the roles that well-placed individuals (e.g., Ban Ki-moon, Gro Harlem Brundtland, Tommy Koh, Mostafa Tolba) have been able to play as leaders who nurture or even ignite processes leading to the diffusion of new normative standards, whether or not these standards appeal to policy-makers located in the most powerful states (Tolba 1998; Brundtland 2005). There is much to be learned about the conditions under which individual leadership can make a difference at the international level (Young 1991). But it would be a mistake to overlook the roles that individuals play, especially with regard to ideational matters like the emergence of ethical principles.

Concluding Thoughts

It is tempting to regard international society as an arena in which power politics reign supreme so that there is little scope for the operation of ethical principles of the sort familiar in other settings. Even among those who accept the relevance of ethical principles at the international level, there is a pronounced tendency to treat such principles as primitive or underdeveloped rules that stand in need of refinement of a sort expected to lead over time to their promotion to the status of legally binding prescriptions setting forth requirements and prohibitions that focus attention on matters of compliance. The account I have developed in this chapter challenges both these assumptions, arguing that each has severe drawbacks as an approach to understanding significant trends unfolding in international society today. No one would argue that a coherent and highly effective system of environmental ethics is currently in place at the international level, or even that such a system is now well along in the process of taking shape. Yet we have come some distance in this realm in recent decades. Precepts like the polluter pays principle, the precautionary principle, and the principle of common but differentiated responsibilities have entered the discourse of world affairs in a manner that is undeniably significant. This does not mean that environmental principles offer simple codes of proper conduct in complex situations of the sort policy-makers regularly confront, much less that the growth of a system of environmental ethics offers unambiguous social capital of use in governing complex systems. As in other social settings, ethical principles achieve influence at the international level by contributing to the framing of issues, providing reasons to act in one way or another, and establishing criteria or benchmarks against which to evaluate performance. They seldom prescribe clear and simple ways to handle specific problems. In this sense, they may well prove helpful in drawing attention to the problems of achieving sustainability in the Anthropocene.

Are we witnessing the development of a meaningful system of international environmental ethics? If so, what vision of human-environment relations lies at the core of this system? We are, at best, at an early stage in the development of international environmental ethics. Although the principles I have discussed are far from trivial, they certainly do not constitute a coherent and effective code of conduct covering the full range of environmental issues arising in international society. Of particular interest, in this connection, is the issue of whether this emerging system is likely to evolve during the foreseeable future as a biocentric ethics of environmental protection or as an anthropocentric ethics of sustainable development. These visions may coexist and even reinforce one another for a period of time. During the foreseeable future, however, the forces leading to an ethical system rooted in the idea of sustainable development are likely to prove stronger than the forces supporting a biocentric system. This is particularly true of the needs for governance characteristic of the Anthropocene. The challenge of coming to terms with the tension between these approaches will become a matter of growing importance over time. But at this stage, the fact that this tension exists and is treated by many as a matter of some importance constitutes evidence that there are issues of substance to be discussed with regard to the development of environmental ethics in international society. The resolution of these issues will play an important role in determining the effectiveness of principled governance as a strategy for dealing with the challenges of the Anthropocene.

7

The Contributions of Good Governance

Introduction

Another perspective that may prove helpful in addressing the challenges of governing complex systems centers on the idea of "good governance." Because this idea is deeply embedded in western political and legal thought, we habitually treat the desirability of pursuing good governance as an article of faith rather than as a matter to be examined critically in analytical and empirical terms. Although understandable, this practice is unfortunate when it comes to assessing the links between good governance and the effectiveness of governance systems, especially in dealing with the complex systems of the Anthropocene. What exactly do we mean in speaking of good governance? Can we differentiate among various elements or dimensions of good governance, and can we devise suitable metrics for measuring the degree to which good governance is present in any given social setting? Critically, can we shed light on the links between good governance and the pursuit of sustainability in the Anthropocene?

Many thoughtful observers take the view that good governance is valuable or desirable in its own right, whether or not it contributes to the effectiveness of governance systems. This position is a perfectly legitimate. In the final analysis, however, a central claim articulated by many (perhaps most) advocates of good governance is that there is a substantial connection between good governance and effective governance. No one would argue that fulfilling the requirements of good governance is sufficient all by itself to guarantee the effectiveness of governance systems. But some do believe that good governance is a necessary condition for effectiveness. And many more espouse the view that good governance can play a significant role in the effort to achieve effective governance (Zaelke et al. 2005).

Most discussions of good governance focus on the establishment and operation of political institutions located at the level of the nation state. We want to know, for example, whether there are opportunities for public participation in the enactment of laws and the development of the regulations needed to implement the laws and whether policy-makers are held accountable for the extent to which their actions conform to key principles of proper conduct on the part of public officials. But there is no reason to limit the analysis of good governance to this level. Although the relevant political institutions are distinctive and prevailing principles of proper conduct differ from case to case, it is perfectly reasonable to ask about good governance in analyzing the performance of traditional institutions operating at the local level. In this chapter, by contrast, I shift the focus of attention in the opposite direction. I am interested in the relationship between good governance and effectiveness with regard to institutional arrangements operating at the level of international society. I turn frequently to the performance of international environmental regimes as a source of illustrations. But my intention is to offer observations about good governance that apply across issue areas and especially about links between good governance and the pursuit of sustainability in the Anthropocene.

International society is anarchical in the sense that there is no government in the ordinary sense of the term in this setting (Bull 1977). But this does not mean that international society lacks political institutions. At the most general level, there are the various elements of the United Nations System. Even more important, I will argue, is the proliferation of international regimes or special purpose governance systems that have grown up to address needs for governance in a variety of issue domains (Krasner 1983; Young 1999). Thus, we now have a range of security regimes (e.g., the nuclear non-proliferation regime), economic regimes (e.g., the world trade regime), and environmental regimes (e.g., the regime for stratospheric ozone). In all these cases, we can ask questions about the links between good governance and the effectiveness of particular governance systems.

Conceptual Issues

Although my examples are predominantly environmental, the critical question is easy to formulate in generic terms. Are there identifiable links between the effectiveness of international regimes and the extent to which they conform to the standards of good governance? Seemingly straightforward when stated in this way, this question turns out to be deceptively simple. A substantial part of the problem has to do with the meaning of the central concepts—effectiveness and good governance.

As chapter 2 makes clear, understanding effectiveness is a challenging task, especially as we move from outputs to impacts lengthening the causal chain and producing a steady increase in the dangers of mistaking spurious relationships for real connections. Nevertheless, we are not without analytic resources relating to effectiveness that are relevant to the theme of this chapter. Some efforts have been made to measure the effectiveness of environmental regimes using a simple nominal scale and applying this scale to a sizable number of regimes (or regime elements) (Miles et al. 2002; Breitmeier, Young, and Zürn 2006; Breitmeier, Underdal, and Young 2011). There are also many qualitative analyses of individual regimes that reflect concerted efforts to address effectiveness in individual cases (Andresen et al. 2012). The conclusions are mixed. But it is probably fair to conclude that most members of the research community believe that regimes can and do make a difference in solving some problems, although there are also a number of other factors that play significant roles in determining both the extent and timing of progress toward problem-solving in international society. Clearly, this situation circumscribes the methods that we can bring to bear on the effort to answer our central question concerning the relationship between good governance and institutional effectiveness. Nevertheless, it certainly does not eliminate the possibility of tackling this question in a serious and sustained manner.

If anything, the conceptual complications surrounding the idea of good governance are harder to untangle. What exactly do we mean in talking about the extent to which governance is good or bad? The answer to this question is anything but clear. Some analysts cast a very wide net in thinking about what makes governance good. One United Nations agency, for example, has suggested that the concept of good governance includes eight distinct elements. Governance is good when it is (i) participatory, (ii) consensus-oriented, (iii) accountable, (iv) transparent, (v) responsive, (vi) effective and efficient, (vii) equitable and inclusive, and (viii) compatible with the rule of law. Another United Nations agency suggests that "[t] o ascertain whether governance is 'good,' actors look at the *mechanisms* that promote it, the *processes* used, and the *outcomes* achieved" (UNDP 2014, 4). This leads to a broad array of specific elements of good governance ranging from the appropriateness of processes to outcomes that are peaceful and supportive of stable and resilient societies.

No doubt, all these factors deserve consideration in the study of governance. But treating good governance as a variable that encompasses such a broad array of distinct dimensions produces a situation that is intractable conceptually and of no real use for purposes of rigorous analysis. For one thing, it absorbs various ideas relating to effectiveness into the definition of good governance, thereby eliminating by definition the central concern of this chapter. For another, it conflates process considerations (e.g., participation, transparency, responsiveness) with outcome considerations (e.g., effectiveness, efficiency, equity, social resilience) in a manner that leaves us with a problem of trying to work with a variable that includes apples and oranges. What is more, the challenge of combining a large number of distinct dimensions into a composite index that is useful in answering a range of key questions is overwhelming. This is not to say that the various aspects of governance that United Nations agencies and others identify are uninteresting or irrelevant. But we need to make some explicit choices to come up with a concept of good governance that is tractable for purposes of analysis as well as helpful in addressing critical questions.

How should we proceed? What I find intriguing and what strikes me as lying at the heart of many discussions of good governance is the idea that processes make a difference when it comes to results arising from the operation of governance systems. That is, the same institutional arrangements can produce different results depending upon the processes involved both in their creation and in their operation or administration. A simple example centers on the observation that people are often prepared to accept regulatory arrangements they have participated in creating that they would reject if these arrangements were imposed in the absence of consultation. The literature on the creation and operation of governance systems contains many similar observations regarding the role of processes. Of course, some processes may be more relevant than others in these terms, and there are different ways to think about similar processes. To make the argument of this chapter tractable, however, I will focus on participation, transparency, and accountability or what I will call the PTA variables.[1] If good governance makes for effective governance, the results should improve as a function of each of the PTA variables. These variables may also interact with one another, but that is a complication we may set aside for the moment in the interests of making progress in addressing the central issue under consideration in this chapter.

Before commenting on the conceptual issues associated with each of the PTA variables, several general observations regarding this way of thinking about good governance are in order. There is great interest, for example, in the extent to which law-making processes at the national level are transparent and the extent to which there are opportunities for public participation in the development of the regulations needed to move the provisions of laws from paper to practice. There is an analogous distinction between the role of good governance in the formation of regimes or governance systems and in the operation or administration of these systems. We can inquire, for instance, whether accountability is more important with regard to the behavior of law makers who may be subject to periodic elections or to the behavior of administrators who may be subject to legislative oversight but who are often able to hide behind an array of bureaucratic defenses.

There is as well an issue regarding the maintenance of good governance over time. An arrangement under which transparency and accountability are high at the outset, for example, may well become opaque, sluggish, and corrupt with the passage of time. None of these observations diminishes the importance of focusing on the PTA variables in this examination of the relationship between good governance and effectiveness with particular reference to steering mechanisms operating in international society. But they do alert us to complications that need to be kept in mind as we think about the significance of the PTA variables.

The idea of participatory democracy has a long and strong pedigree (Dryzek 1990). The central claim is easy to grasp. Citizens not only have a right to participate in making decisions about the governance systems guiding their behavior; they are also more likely to regard the results as legitimate and to comply with relevant requirements and prohibitions if they feel their voices are heard in the creation and administration of these arrangements. This is the basic rationale underlying a vigorous defense of the right to vote in fair elections in democratic systems. Leaving aside the complexities of this idea for the moment, it is also worth noting that we have seen in recent times a sharp rise in interest in providing opportunities for public participation in the development of regulations needed to implement legislation, and in the use of these regulations to address issues on a case-by-case basis. Nowhere is this more apparent than in efforts to address environmental issues. The case of public participation under the provisions of the 1969 National Environmental Policy Act (NEPA) in the United States provides a dramatic example of this development. But similar processes are in evidence in other settings.

Yet public participation is a complex phenomenon that does not lend itself to easy measurement. Part of this has to do with issues relating to the identification of those whose voices should be heard in various settings. It has become fashionable to speak of stakeholders in many settings and to proclaim that it is essential to listen to the concerns of stakeholders. But an examination of actual practices in this realm suggests that there is seldom any rigorous way to determine who the stakeholders are and that many efforts to consult stakeholders are pro forma or relatively empty exercises. In addition, there are issues surrounding what the International Association of Public Participation calls the Spectrum of Public Participation (see figure 7.1).[2] What the idea of the spectrum makes clear is that the notion of public participation covers a multitude of related but differentiable processes ranging from keeping stakeholders informed to placing actual decision-making in the hands of the public. In between, we can identify a wide range of procedures including public comments, focus groups, citizen advisory committees, and so forth. In each case, there is great variation in how these procedures work in practice. What leads to meaningful participation in one setting may atrophy into a pro forma exercise that is largely irrelevant in another setting.

Curiously, the analogue to public participation at the international level may be more straightforward. The members of international environmental regimes are for the most part sovereign states (in contrast to individual citizens in domestic society) that cannot be bound legally and will not be bound politically without their consent. Each state that has

IAP2 Spectrum
of Public Participation

International Association
for Public Participation

Increasing *Level of Public Impact*

	Inform	Consult	Involve	Collaborate	Empower
Public participation goal	To provide the public with balanced and objective information to assist them in understanding the problem, alternatives, opportunities and/or solutions.	To obtain public feedback on analysis, alternatives and/or decisions.	To work directly with the public throughout the process to ensure that public concerns and aspirations are consistently understood and considered.	To partner with the public in each aspect of the decision including the development of alternatives and the identification of the preferred solution.	To place final decision-making in the hands of the public.
Promise to the public	We will keep you informed.	We will keep you informed, listen to and acknowledge concerns and aspirations, and provide feedback on how public input influenced the decision.	We will work with you to ensure that your concerns and aspirations are directly reflected in the alternatives developed and provide feedback on how public input influenced the decision.	We will look to you for advice and innovation in formulating solutions and incorporate your advice and recommendations into the decisions to the maximum extent possible.	We will implement what you decide.
Example techniques	▪ Fact sheets ▪ Web sites ▪ Open houses	▪ Public comment ▪ Focus groups ▪ Surveys ▪ Public meetings	▪ Workshops ▪ Deliberative polling	▪ Citizen advisory committees ▪ Consensus-building ▪ Participatory decision-making	▪ Citizen juries ▪ Ballots ▪ Delegated decision

Figure 7.1

acceded to the 1987 Montreal Protocol on Substances that Deplete the Ozone Layer, for example, has done so on the basis of its own calculations regarding the merits of joining the ozone regime. The obligations that each member has assumed have become legally binding on the basis of the procedures governing the ratification of international agreements within individual member states. In the case of the United States, which is famous for the stringency of its ratification procedures, an agreement like the Montreal Protocol does not become binding until it is ratified by a two-thirds majority of the Senate. In some states, there is also a need to enact implementing legislation to provide responsible agencies with the mandate and the authority they need to promulgate suitable regulations and to administer them on a day-to-day basis. What is more, when it comes to the operation of an arrangement like the Montreal Protocol, member states are able to participate actively in the deliberations of the Conference/Meeting of the Parties. In most cases, decisions of the COP/MOP are taken by consensus, an arrangement that minimizes the likelihood that major developments affecting the operation of the regime will take place in the absence of consent on the part of individual members.[3] Of course, the extent to which there are opportunities for public participation *within* member states regarding their engagement in international regimes will vary from state to state; some observers view the low level of public participation occurring within some member states as an obstacle to be overcome in the effort to make international regimes effective as tools for problem-solving. Nevertheless, it is hard to argue with the proposition that member states themselves normally enjoy a high level of participation in the creation and operation of international regimes.

Transparency is the second of the PTA variables. Transparency is a measure of the extent to which the creation and operation of governance systems feature open processes that allow all those interested to see how rules are formulated and applied to specific situations. Transparency is widely touted as an important element of good governance. But this does not mean that actual practices are transparent, especially in cases where the issues are complex and there are multiple opportunities for interested parties to exercise influence behind the scenes. Legislative processes that are open to the public in formal terms, for example, often feature bargaining that takes place behind closed doors producing backroom deals in which lobbyists for various interest groups participate actively. In cases where legislation involves agreement on the provisions of lengthy and highly technical documents, it is extremely difficult to fulfill the ideal of transparency. In the United States, for example, acts of Congress now routinely run to hundreds or even thousands of pages. Even congressional staff members—much less the members of Congress themselves—are often unable to scrutinize and assess the significance of provisions inserted in bills by various participants in the legislative process.

If anything, transparency is even harder to achieve in the bureaucratic settings that become the focus of attention when we get to the stage of implementing the provisions of governance systems. Despite the growing emphasis on public participation in these settings, few people are able to understand the ins and outs of the bureaucratic processes involved in the administration of complex governance systems. Even the bureaucrats themselves may have a hard time following the administration of the different elements of complex systems. Under the circumstances, bureaucracy can degenerate into a labyrinthine and opaque jungle that has little connection to solving societal problems.

In response to these concerns, an organization named Transparency International (TI) has taken up the challenge of evaluating the status of transparency in the governance practices of nation states. TI has worked hard to devise indicators that allow us to get a sense of the status of this element of good governance and to compare countries in terms of transparency. The results are not straightforward. But perhaps the most striking component of this effort features what TI calls the Corruption Perceptions Index. This index ranks countries on a scale from 0 to 100 where 0 represents the maximum level of perceived corruption and 100 represents the maximum level of cleanliness. The results, displayed in figure 7.2 for 2013, are not particularly surprising.[4] The Nordic countries and other small countries like New Zealand, Singapore, and Switzerland are outstanding performers in these terms. The usual suspects in Africa and Asia are scored as highly corrupt. It is interesting to note that India is regarded as more corrupt than China and that Russia comes in at number 127 out of the 177 countries included in the ranking. Of course, this index focuses on "perceived" levels of public sector corruption. But perceptions of corruption are important in their own right, and it is probable that such perceptions would correlate well with any of a variety of other indicators of corruption or transparency that might seem to be more objective.

TRANSPARENCY INTERNATIONAL
the global coalition against corruption

CORRUPTION PERCEPTIONS INDEX 2013

The perceived levels of public sector corruption in 177 countries/territories around the world.

SCORE

Highly Corrupt 0-9 10-19 20-29 30-39 40-49 50-59 60-69 70-79 80-89 90-100 Very Clean   No data

#stopthecorrupt
www.transparency.org/cpi

RANK	COUNTRY/TERRITORY	SCORE
1	Denmark	91
1	New Zealand	91
3	Finland	89
3	Sweden	89
5	Norway	86
5	Singapore	86
7	Switzerland	85
8	Netherlands	83
9	Australia	81
9	Canada	81
11	Luxembourg	80
12	Germany	78
12	Iceland	78
14	United Kingdom	76
15	Barbados	75
15	Belgium	75
15	Hong Kong	75
18	Japan	74
19	United States	73
19	Uruguay	73
21	Ireland	72
22	Bahamas	71
22	Chile	71
22	France	71
22	Saint Lucia	71
26	Austria	69
26	United Arab Emirates	69
28	Estonia	68
28	Qatar	68
30	Botswana	64
31	Bhutan	63
31	Cyprus	63
33	Portugal	62
33	Puerto Rico	62
33	Saint Vincent and the Grenadines	62
36	Israel	61
36	Taiwan	61
38	Brunei	60
38	Poland	60
40	Spain	59
41	Cape Verde	58
41	Dominica	58
43	Lithuania	57
43	Slovenia	57
45	Malta	56
46	Korea (South)	55
47	Hungary	54
47	Seychelles	54
49	Costa Rica	53
49	Latvia	53
49	Rwanda	53
52	Mauritius	52
53	Malaysia	50
53	Turkey	50
55	Georgia	49
55	Lesotho	49
57	Bahrain	48
57	Croatia	48
57	Czech Republic	48
57	Namibia	48
61	Oman	47
61	Slovakia	47
63	Cuba	46
63	Ghana	46
63	Saudi Arabia	46
66	Jordan	45
67	Macedonia (FYR)	44
67	Montenegro	44
69	Italy	43
69	Kuwait	43
69	Romania	43
72	Bosnia and Herzegovina	42
72	Brazil	42
72	Sao Tome and Principe	42
72	Serbia	42
72	South Africa	42
77	Bulgaria	41
77	Senegal	41
77	Tunisia	41
80	China	40
80	Greece	40
82	Swaziland	39
83	Burkina Faso	38
83	El Salvador	38
83	Jamaica	38
83	Liberia	38
83	Mongolia	38
83	Peru	38
83	Trinidad and Tobago	38
83	Zambia	38
91	Malawi	37
91	Morocco	37
91	Sri Lanka	37
94	Armenia	36
94	Benin	36
94	Colombia	36
94	Djibouti	36
94	India	36
94	Philippines	36
94	Suriname	36
102	Ecuador	35
102	Moldova	35
102	Panama	35
102	Thailand	35
106	Argentina	34
106	Bolivia	34
106	Gabon	34
106	Mexico	34
106	Niger	34
111	Ethiopia	33
111	Kosovo	33
111	Tanzania	33
114	Egypt	32
114	Indonesia	32
116	Albania	31
116	Nepal	31
116	Vietnam	31
119	Mauritania	30
119	Mozambique	30
119	Sierra Leone	30
119	Timor-Leste	30
123	Belarus	29
123	Dominican Republic	29
123	Guatemala	29
123	Togo	29
127	Azerbaijan	28
127	Comoros	28
127	Gambia	28
127	Lebanon	28
127	Madagascar	28
127	Mali	28
127	Nicaragua	28
127	Pakistan	28
127	Russia	28
136	Bangladesh	27
136	Côte d'Ivoire	27
136	Guyana	27
136	Kenya	27
140	Honduras	26
140	Kazakhstan	26
140	Laos	26
140	Uganda	26
144	Cameroon	25
144	Central African Republic	25
144	Iran	25
144	Nigeria	25
144	Papua New Guinea	25
144	Ukraine	25
150	Guinea	24
150	Kyrgyzstan	24
150	Paraguay	24
153	Angola	23
154	Congo Republic	22
154	Democratic Republic of the Congo	22
154	Tajikistan	22
157	Burundi	21
157	Myanmar	21
157	Zimbabwe	21
160	Cambodia	20
160	Eritrea	20
160	Venezuela	20
163	Chad	19
163	Equatorial Guinea	19
163	Guinea-Bissau	19
163	Haiti	19
167	Yemen	18
168	Syria	17
168	Turkmenistan	17
168	Uzbekistan	17
171	Iraq	16
172	Libya	15
173	South Sudan	14
174	Sudan	11
175	Afghanistan	8
175	Korea (North)	8
175	Somalia	8

Figure 7.2

Again, most efforts to address issues of transparency focus on the level of the nation state rather than the level of international governance. Thinking about transparency at the level of international governance introduces several additional issues. International negotiations often proceed in closed sessions (Chasek and Wagner 2012). They are highly opaque from the perspective of members of the general public; they may even involve secret agreements that do not become known to the public until much later. Treating states as the relevant actors, on the other hand, it seems fair to say that there is a relatively high level of transparency in the intergovernmental processes involved in creating international regimes. Many member states complained about the lack of transparency in the process leading to the introduction of the text of the Copenhagen Accord at the eleventh hour of COP 15 of the UN Framework Convention on Climate Change in December 2009. But this was a highly unusual case, and the lack of transparency in the process of formulating the accord goes far toward explaining the facts that in the end the COP took note of the accord but did not adopt it formally, and that those in charge of climate negotiations have made a point of avoiding such procedures since Copenhagen. Those managing the negotiations leading to the Paris Agreement at COP 21 in December 2015, for example, made a concerted and apparently successful effort to avoid problems of transparency. In the ordinary case, then, the parties are fully informed about the proposed terms of international agreements during the negotiation stage.

When it comes to implementation, much of the action shifts to the level of the member states. The administration of international regimes does involve periodic conferences of the parties and the activities of small secretariats (Jinnah 2014). For the most part, what goes on in these settings is fairly transparent. But implementation at the national level is a different story. There is great variation among the members of most international regimes regarding the efforts they make to implement the provisions of these arrangements, much less the degree to which their efforts are transparent. Some public agencies are more forthcoming about their activities than others, and some states have oversight procedures (e.g., congressional oversight in the US) that can enhance transparency in this context. In general, however, transparency regarding the implementation of international regimes is a matter regarding administrative practices at the national level that are often opaque.

The third of the PTA variables is accountability. How should we think about accountability with regard to the effectiveness of governance systems and especially issue-specific regimes? From the perspective of democratic theory, accountability is a matter of legislative control over the actions of the executive and popular control over the actions of the legislature via periodic elections. Legislators must defend their voting records when they stand for reelection. For its part, the executive is not only accountable to members of the electorate but also to the legislature which has the power to alter the authority granted to the executive under various laws and to control the funds available to the executive to act in particular ways. In the case of international regimes, agreements negotiated and signed by the executive often require ratification on the part of the legislature and may require the passage of implementing legislation. In the United States, however, there are prominent cases (e.g., the UN Convention on the Law of the Sea) in which the executive has participated in negotiating an agreement that the Senate has subsequently refused to ratify, preventing the agreement from becoming legally binding on the United States.[5]

It does not take a lot of analysis, however, to see that there are serious problems with this conception of accountability, at least with regard to the creation and implementation of regimes dealing with specific issue domains. Elections are sensitive to many issues; they seldom have much to do with the actions of legislators or chief executives relating to the creation or operation of specific international regimes. In those rare cases where issues relating to specific regimes do become visible in electoral settings (e.g., the reception of the 1919 Treaty of Versailles in the US), the influence of well-funded special interest groups is often so great that it is questionable whether elections play a significant role with regard to accountability. Similar comments are in order regarding legislative oversight of the activities of executive agencies in the realm of implementation. Except in unusual cases that trigger substantial public interest, serious oversight processes are uncommon. Moreover, it is routine for alliances to develop between responsible administrative agencies and the legislative (sub)committees that are charged with making use of oversight procedures to ensure that agencies are held accountable for their actions in specific issue domains. This is not to say that there is no accountability regarding the creation and administration of regimes dealing with specific issues. But it is clear that we cannot count on high levels of accountability, even in democratic settings. Of course, the whole issue of accountability may be moot in authoritarian settings where there are no effective processes in place to ensure public accountability of government actions in general.

Somewhat surprisingly, accountability may be stronger with regard to international regimes than it is in the case of ordinary domestic arrangements.[6] This is in part a function of the requirement for ratification in the case of international agreements. In the United States, for example, the prospect that the Senate will refuse to ratify a convention or a treaty is sufficiently real to ensure that the executive will pay careful attention to the concerns of senators during the process of negotiating an agreement. It is common to include senators in American delegations to major negotiations, and the provisions of prospective agreements are often vetted with key senators and members of their staffs during the negotiation process. Still, it would be a mistake to make too much of this form of accountability. Many states have ratification procedures that are far less stringent than those of the United States. Presidents and prime ministers are often able to employ executive agreements as a means of bypassing the normal requirements for ratification.[7] Even in cases where the legislature has refused repeatedly to ratify an agreement (e.g., the Law of the Sea Convention in the US), key provisions may acquire the status of customary international law with the passage of time, a status that makes them legally binding even in the absence of formal ratification.

When it comes to the administration of international arrangements that have entered into force, processes designed to ensure accountability are apt to be particularly ineffective. In many cases, governments fail to comply with the reporting requirements embedded in international regimes, much less the substantive provisions of these arrangements (Brown Weiss and Jacobson 1998; Victor, Raustiala, and Skolnikoff 1998). More often than not, there are no real efforts to hold governments accountable in these terms. Occasionally, bodies like the US Government Accountability Office (GAO) issue reports noting the lackadaisical manner in which such requirements are addressed. But even then, there is apt to be no systematic or sustained response on the part of the relevant agencies.

Where does this leave us regarding the central question of this chapter? It turns out that it is easier to make casual statements regarding the nature and importance of good governance than it is to turn this concept into a criterion that is usable for purposes of rigorous analysis. I am particularly interested in what good governance has to do with the creation and administration of regimes dealing with various more-or-less specific issue domains and particularly in the role of what I have labeled the PTA variables. This allows us to ask intuitively appealing questions regarding the extent to which participation, transparency, and accountability are linked to the success of governance systems in solving societal problems. But even here complications abound. We do not have any straightforward way to measure these variables within a particular political system, much less in more generic terms. It turns out that what may seem simple at the conceptual level becomes quite convoluted when we get down to the details of specific cases. In this regard, it is interesting to observe that operationalizing the PTA variables may be more straightforward in dealing with international regimes than in dealing with mainstream governance systems at the domestic level. Under international law, states cannot be bound legally in the absence of their explicit consent, a factor that ensures that individual states have numerous opportunities to participate in the development of the provisions of international agreements. Transparency is also relatively high; cases like the Copenhagen Accord in which parties do not have access to the terms of proposed agreements well in advance are rare. And the requirements regarding ratification at the national level put pressure on those negotiating on behalf of individual states to be responsive to the concerns of those whose votes are needed during the ratification process. Does this mean we can count on a high level of good governance with regard to the creation and operation of international regimes? I am skeptical about offering an affirmative answer to this question. But it does give us something to go on in moving from the level of conceptual issues to the level of theoretical matters.

Theoretical Matters

What I have said so far is largely a matter of ground clearing. It constitutes an effort to clarify variables sufficiently to make it possible to formulate and evaluate hypotheses concerning the relationships between or among the key variables. Of course, the basic concern is simple. We want to analyze the widespread belief, at least in the realm of liberal democratic thought, that good governance is a significant source of effective governance. But we need to push the argument beyond this simple connection. What is the nature of the links between the effectiveness of governance systems and the PTA variables? Is one of these variables more important than the others? Do they interact with one another as sources of effectiveness? Above all, can we identify the mechanisms through which the PTA variables are thought to contribute to the effectiveness of governance systems? To produce usable knowledge, it is not enough to determine whether there are correlations between effectiveness and the PTA variables. We need to understand the causal connections associated with these relationships in order to take well-informed steps to enhance the effectiveness of governance systems operating in specific settings and particularly in the setting we now think of as the Anthropocene. In this section, I explore these matters, paying particular attention to ideas concerning the mechanisms involved.

Several clarifying observations will help to set the stage for this exercise. What is important in determining the success of regimes in solving problems once they are in place may differ from the factors that loom large in the process of creating regimes in the first place. My chief concern here is the operation of regimes, though I will extend the discussion to issues of regime formation whenever it seems helpful to do so. In addition, it is important to be on the alert for scope conditions that may be important as determinants of the relationship between the PTA variables and effectiveness. That is, the various elements of good governance may contribute to effectiveness under some conditions but not under others. What is important in democratic political settings where people expect to be provided with opportunities to participate, for example, may be much less important in more authoritarian settings where there are no entrenched expectations in this regard.

The role of participation in promoting the effectiveness of governance systems rests on an observation that many analysts have made in a variety of social settings. People who feel they have had an opportunity to participate in a meaningful fashion in determining the content of the rules of the game are not only willing to agree to rules they would otherwise find unacceptable; they are also more likely to comply with the relevant requirements and prohibitions voluntarily or without reference to the prospect of being penalized for non-compliance. An interesting feature of this observation is that it appears to run counter to a central premise of utilitarian thinking (Thaler 2015). If the benefits of compliance outweigh the costs, utilitarian thinking suggests that subjects ought to opt for compliance whether or not they have participated in some meaningful way in establishing the relevant requirements and prohibitions. Conversely, if the benefits of non-compliance outweigh the costs, subjects should refuse to comply with the requirements and prohibitions whether or not they have participated in the process of establishing them. Of course, a determined utilitarian can devise an argument to save the utilitarian view of things in this context. Such an argument might stress the benefits to be derived from participation as an end in itself or the broader benefits associated with a system that conforms to some norm of self-governance. But it may make more sense in this context to follow the thinking of the social psychologists whose work underlies what we now think of as behavioral economics (Ariely 2010; Kahneman 2011; Thaler 2015). There are limits to the power of arguments framed in terms of the role of calculations of benefits and costs as determinants of behavior. People simply may react differently to requirements and prohibitions when they have a sense that they have participated in some meaningful way in establishing the rules rather than being subjected to rules imposed on them by others.

Needless to say, there may be a cultural bias embedded in this argument. Public participation, which is regarded as a great virtue in liberal democratic thought, is not a central concern regarding governance in all political settings. Even in democratic systems featuring many opportunities for public participation, policy-makers allocate substantial resources to the development and deployment of enforcement systems capable of imposing or threatening to impose sanctions on those who violate legally binding prescriptions (Paddock et al. 2012). And there is little evidence to suggest that voluntary compliance is weaker in more authoritarian systems in which the rules are imposed from above than it is in liberal democratic systems. What is more, the focus on compliance deflects attention from what students of effectiveness characterize as impacts.[8] There are numerous cases in which high levels of compliance with the rules of the game are not sufficient to solve the problem with which they are associated. In fact, there is some danger that focusing a great deal of attention on compliance will turn out to be a distraction with regard to the underlying concern for solving major societal problems.

It is worth noting as well that the relationship between participation and effectiveness may differ at the international level. Given the central role of consensus at this level, states are seldom required to comply with rules that they have not participated in formulating; implementation is left for the most part for member states to deal with on their own. In a sense, therefore, we can say that there is a high level of participation in most international governance systems. But this hardly ensures that member states will comply with the terms of these arrangements, much less be motivated to do so due to a sense of having had a say in the development of the rules. Small or weak states often feel that they have been pressured into accepting the terms of a regime, even when they accept the terms formally (Gruber 2000). States are also collective entities, a fact that can complicate compliance with the provisions of international regimes greatly. Those agencies assigned to implement the provisions of a regime may or may not have the capacity or the will to handle such assignments effectively, even in cases where an agreement has been ratified and is legally binding on the state as a whole. Above all, states often lack a continuing sense of obligation. It is by no means uncommon for new administrations to renounce commitments entered into by their predecessors; even the same government may decide that it is no longer expedient to make a concerted effort to comply with obligations acquired voluntarily at some earlier stage in its tenure.[9] These factors suggest that we should be cautious in transferring arguments about the relationship between participation and effectiveness derived from observations relating to the behavior of individuals to the analysis of effectiveness regarding international regimes.

What about transparency? The logic underlying the idea that transparency enhances the effectiveness of governance systems bears some similarity to the argument regarding participation. Once again, the focus is largely on compliance: those who know that their behavior will be monitored closely and publicized widely are thought to be less likely to violate the rules than those who believe that their non-compliant behavior will remain hidden. There are at least two distinct mechanisms that can lead to such a relationship. One is largely utilitarian. Those whose non-compliant behavior becomes a matter of public record may be influenced by the expected costs involved, either in the form of formal sanctions imposed by some public authority or in the form of more informal but nevertheless significant sanctions involving reputational effects. Those who look at the big picture may well regard the cost of complying with the prescriptions of specific regimes as modest compared to the costs associated with acquiring a reputation for unreliability that limits their ability to enter into beneficial agreements in other domains or at other times. The other mechanism is more cultural in nature. There are those who are willing to violate rules so long as they are able to do so clandestinely but who are much less likely to do so when their behavior becomes a matter of public record and especially when it becomes known to others whose opinions they value highly. At the individual level, this mechanism is often associated with the idea of shame in contrast to guilt. Anthropologists and sociologists, for example, speak of shame cultures in which clandestine violations are acceptable but in which there is more or less extreme shame associated with such behavior becoming known to other members of the community (Benedict 2006).

Here, too, it is worth asking whether the relationship between transparency and effectiveness based largely on observations of individual behavior carries over into the realm of international regimes. There is, to begin with, an issue regarding the nature of transparency at the international level. States seldom acknowledge that they are not complying with the terms of international agreements, and there are many cases in which it is not easy to determine with any precision whether individual member states are or are not violating the terms of an agreement. But setting this complication aside for the moment, what can we say about the role of transparency at the international level? It is difficult to identify an analogue of shame when it comes to accounting for the behavior of collective entities like states. But the leaders of states certainly do pay attention to reputational issues (Tomz 2007). A state that acquires a reputation for being unreliable when it comes to living up to commitments or formal obligations accepted under the terms of international agreements may find its options severely constrained when it comes to negotiating the terms of new agreements dealing with a variety of important issues. At the international level, moreover, capacity is an important consideration. Some states simply lack the capacity to comply with provisions requiring more or less substantial adjustments to existing conditions (Chayes and Chayes 1995). This is what underlies both the principle of common but differentiated responsibilities embedded in the provisions of many international environmental regimes and the understanding that financial assistance is needed to allow less developed countries to participate effectively in some regimes. This explains the distinction regarding the obligations of UNFCCC Annex 1 and non-Annex 1 countries under the terms of the 1997 Kyoto Protocol as well as the creation of the Montreal Protocol Multilateral Fund to assist developing countries in efforts to avoid or phase out the production and consumption of ozone-depleting substances.

The role of accountability in making governance systems effective differs somewhat from the roles of participation and transparency. Whereas the variables P and T are focused on the behavior of those subject to a regime's prescriptions, the idea of accountability reflects the fact that in many situations the focus is on the actions of agents authorized to act on behalf of subjects rather than on the behavior of the subjects themselves. This is an important consideration in most political settings. We elect officials to act on our behalf and entrust them with the responsibility of making a good faith effort to carry out our wishes, including complying with the requirements and prohibitions of various governance systems. In large-scale systems, elected officials appoint a range of civil servants to carry out the tasks that have been assigned to them. At the local level, this is apt to involve functions like the maintenance of public order and the provision of public education. At the national level, it encompasses a wide range of assignments, including fulfilling obligations set forth in the terms of international environmental agreements. Of course, we hope that both elected officials and the civil servants they appoint will be upstanding individuals who are motivated to carry out their responsibilities even in the absence of social pressures to do so. Nevertheless, we cannot rely on such motivations to ensure that these individuals will always or even most of the time act in a way that justifies such hopes.

There are certainly variations across social settings and time periods regarding such matters. But all societies confront problems of accountability on a regular basis. In democratic systems, the assumption is that the need to stand for reelection from time to time will ensure that elected officials are accountable for their actions and that they in turn will provide the discipline needed to ensure that civil servants are held accountable. But experience makes it clear that there is great variation in these terms. Elected officials can and often do interpret accountability in a lax manner; civil servants are often in a position to ignore the requirements of accountability even while paying lip service to them. Both opportunities and incentives to engage in corrupt practices are great in many situations, especially where there is room to influence the collection and expenditure of public funds. This is true even in political systems that are generally regarded as democratic.[10] Needless to say, the limits of accountability are even more severe in many political systems that are less democratic in character. All this simply reinforces the argument about the importance of accountability as a determinant of effectiveness with regard to the provisions of governance systems. The greater the temptations for key actors to act in ways that detract from efforts to solve societal problems or even to undermine such efforts, the more important is the existence of accountability mechanisms in ensuring that governance systems are successful in solving problems.

The issue of accountability with regard to the provisions of international regimes raises questions at two levels. For the most part, international administration is thin on the ground and limited in terms of its capacity. But it is not non-existent (Biermann and Siebenhüner 2009; Jinnah 2014). It does make a difference whether the secretariats established under the terms of international environmental agreements carry out their assigned tasks energetically and faithfully and whether those responsible for managing international financial mechanisms (e.g., the ozone regime's Montreal Protocol Multilateral Fund or the climate regime's Clean Development Mechanism) are upstanding individuals. But because much of the responsibility for the implementation of the provisions of international regimes falls to agencies operating within member states, a great deal of the concern about accountability focuses on the behavior of these agencies. There is no reason to believe that such agencies are more or less lax in carrying out obligations derived from international agreements than obligations embedded in domestic legislation. There are even cases in which responsible agencies become vigorous advocates for the regimes they serve in interactions with other agencies within the same government or with politicians who are skeptical or even opposed to fulfilling obligations embedded in international agreements (Dai 2005; Dai 2007). But this does not alter the fact that much of the concern about accountability in conjunction with international regimes is directed to the actions of agents of national governments rather than agents associated with international administrative bodies.

The PTA variables are not so tightly coupled that they cannot operate as forces promoting the effectiveness of governance systems on an individual basis. On the contrary, each of these variables has its own logic, and the argument regarding the role of each has evolved on its own terms. Participation may make a difference as a driver of compliance, regardless of the level of transparency or accountability. The importance of transparency does not depend on opportunities for subjects to participate in the development of requirements and prohibitions. Accountability is clearly, perhaps especially, relevant in cases marked by low levels of participation and transparency.

Even so, it is relevant to ask whether there are synergies among the PTA variables that can strengthen the role that these process considerations play in enhancing the success of governance systems in solving problems. Perhaps the most obvious case involves the interaction between transparency and accountability. Surely, it is easier to hold both policy-makers and civil servants accountable when there are good reasons to expect that their behavior will become a matter of public knowledge. It may be as well that the PTA variables form an influential cluster when they operate together to fulfill our ideal of good governance. According to this logic, when subjects feel they have participated meaningfully in making the rules, that the behavior of all relevant parties will become known to members of the attentive public, and that the operation of accountability mechanisms ensures that agents make a good faith effort to fulfill their obligations, the prospects for solving a wide range of societal problems, including the newly emerging problems of governing complex systems in the Anthropocene, will rise.

Empirical Observations

What, then, do we know about the empirical connections between the PTA variables and the effectiveness of governance systems, and international regimes more specifically? What opportunities exist to improve our understanding of these links? The bottom line regarding these questions is straightforward. Our current knowledge regarding these links is limited. There is not much we can say with great confidence about the role of good governance in enhancing the effectiveness of governance systems as mechanisms for solving a variety of societal problems, much less for the pursuit of sustainability in the Anthropocene. On the other hand, it may well be that there is room for launching new research efforts that could add significantly to our understanding of these links. Because we are starting from an extremely modest point of departure in this realm, most any serious effort to explore the links empirically will constitute a significant addition to the state of knowledge in this realm. And because support for the ideal of good governance is so widespread, any work that sheds light on this relationship should be of interest to a sizable audience, including especially those whose point of departure is an interest in the role of process considerations like the PTA variables. In this section, I provide preliminary answers to these questions, beginning with what we know already and moving to what we can realistically hope to find out.

The first hurdle in empirical terms is the lack of a tractable and widely accepted metric to use in assigning scores to the effectiveness of international regimes, much less to the effectiveness of governance systems more generally. Recalling the general account of effectiveness in chapter 2, we can highlight several issues of particular importance to this analysis of the relationship between good governance and effectiveness. The challenge is to find ways to compare actual performance (AP) with both the no-regime counterfactual (NR) and the collective optimum (CO). At this stage, we lack straightforward and widely accepted procedures for assigning values to these variables in real-world cases (Young 2003). Some analysts have sought to simplify the procedure, focusing only on actual performance rather than including some measure of progress toward achieving the collective optimum (Breitmeier, Underdal, and Young 2011). But this still leaves us with the difficult and controversial task of coming to terms with the counterfactual. Would the problem have remained unsolved or been solved through other means in the absence of the relevant regime?

There are, on the other hand, many studies that tackle the question of effectiveness through detailed studies of individual cases. An interesting observation in this regard is that many political scientists have a tendency to direct attention to outputs in contrast to outcomes and impacts. They take an intense interest in the processes leading to the enactment of legislation and remain concerned with issues of implementation (e.g., the promulgation of regulations) arising in making the transition from paper to practice. But they often lose interest once the process moves on from there. Among those who do think about outcomes and impacts, there is a divide between those (often political scientists) who focus on matters of compliance on the part of subjects and those (often economists) who direct attention to various measures of impacts, like reductions in emissions of sulfur dioxide or reductions in the production and consumption of ozone-depleting substances. In cases where both groups of analysts have looked at the same international regime (e.g., the regime dealing with transboundary air pollution in Europe), there is often a substantial difference in findings regarding effectiveness. The behavioralists typically find more evidence of the effectiveness of regimes than those who concentrate on determining the extent to which regimes contribute to solving or at least ameliorating the relevant problems.

Be that as it may, there is widespread agreement that some international regimes do matter at least some of the time. Research dealing with a wide range of individual cases has suggested a number of factors that seem to contribute to effectiveness (Young 2011a; Chapter 2 *infra*). Some findings are general propositions like the conclusion that the presence of a single dominant actor (what political scientists typically call a hegemon) is not a necessary condition for effectiveness. Others are context specific like the proposition that effective regimes generally reflect a careful effort to match the attributes of the regime to the most important properties or features of the problem at hand. These studies have not focused systematically on relationships between values of the PTA variables and observed levels of effectiveness. Perhaps the most relevant finding is that the development and maintenance of feelings of fairness or legitimacy make a difference, especially in cases where regimes are expected to operate on a continuous or indefinite basis. Although neither fairness nor legitimacy is reflected in any direct way in the PTA variables, it is plausible to argue that there is some connection between participation and the feeling that the provisions of a regime are fair or that accountability is a necessary (though not sufficient) condition for the maintenance of a sense that a regime meets some standard of legitimacy.

Focusing specifically on the PTA variables, somewhat similar observations are in order. In the case of participation, there are questions about who is allowed to participate, what form participation takes, and whether participation extends to the stage of day-to-day operation or is limited to the stage of regime formation. As I observed earlier, there is a sense in which international regimes constitute a special case in these terms. If we treat member states as the subjects of such regimes, participation is unusually high. Of course, individual states may feel pressured or coerced into accepting the preferences of the majority or of a particularly powerful state. Even so, no state is bound to accept the terms of any international regime against its will. For the most part, member states also are not required to accept decisions made under the auspices of international regimes against their will. There are exceptions to this generalization. The International Whaling Commission has the authority to make some important decisions by a three-fourths majority. The Meeting of the Parties to the Montreal Protocol can make decisions about matters like the acceleration of phase out schedules for ozone-depleting substances already covered that take effect without requiring ratification on the part of individual member states. But even in these cases, the prevalence of consensus practices and the option of filing an objection or reservation in extreme cases create a situation in which the opportunity for individual members to participate in decisions regarding the operation of regimes is unusually high (Breitmeier, Young, and Zürn 2006).

Is there any evidence, that these practices relating to participation are linked to the effectiveness of international regimes? Certainly, this form of participation is not sufficient to ensure that regimes will prove effective. There are numerous cases of regimes that perform poorly or fail outright despite providing ample opportunities for subjects to participate in their operation. On the other hand, it is possible to make a case for the proposition that high levels of participation are necessary (though not sufficient) to the success of international regimes. This may seem like an effort to make a virtue out of a necessity. In a society of sovereign states, there is no alternative to arrangements in which regime members enjoy a high level of participation regarding the development and the administration of the rules of the game. Nonetheless, it is difficult to escape the conclusion that high levels of participation are linked in some way to success. The opportunity to participate may well play a role in the thinking of those responsible for ensuring that member states comply with the obligations they accept as signatories to the agreements establishing international regimes. This is a significant, empirically grounded example of a likely connection between good governance and effectiveness with regard to international regimes.

Transparency turns out to be a major problem in connection with the operation of international regimes. There are often ways to conceal or divert attention from behavior that does not conform to the obligations of regime membership. Given the problems of moving from correlations to causal connections regarding environmental issues, it may be difficult to assign blame to specific parties, even when there is clear evidence regarding such matters as the origins of waterborne and airborne pollutants. In many cases, member states are remarkably lax even in meeting simple reporting requirements included in the provisions of international agreements. This combination of conditions is what has given rise to controversies regarding what is often referred to as monitoring, reporting, and verification (MRV) in connection with the operation of specific regimes. Although participants often look for a high level of transparency regarding the behavior of individual regime members, the frequent controversies associated with issues pertaining to MRV attest to the fact that such transparency is often hard to come by in concrete situations.

What is the relationship between transparency in this sense and the effectiveness of international regimes? The answer to this question may be related to the impact of violations on the performance of regimes. When even a single violation of a regime's obligations (e.g., the conduct by a member of a nuclear test in violation of the terms of the 1996 Comprehensive Nuclear-Test-Ban Treaty) may cause the collapse of a regime, transparency is likely to be critical. That is, the success of the regime depends on the operation of MRV procedures that eliminate or at least minimize the possibility of clandestine violations. This is why transparency is a matter of extreme sensitivity in the context of most arms control regimes (Brennan 1961). Other cases are likely to be less sensitive to occasional violations or low levels of violation. There is a black market in ozone-depleting substances, for example, that does not pose a serious threat to the effectiveness of the ozone regime or engender conditions leading important members to consider withdrawing from participation in the regime. This is a case that illustrates my earlier remark about scope conditions. Transparency, exemplified in the operation of sophisticated MRV procedures, seems to be a key determinant of regime effectiveness under some conditions but not under others. The challenge for research, then, is to bear down on the effort to sharpen our understanding of the nature and content of the relevant scope conditions.

The role of accountability in the context of international regimes is in some ways opposite to the role of participation. The secretariats created to administer certain provisions of these regimes and to provide administrative services for meetings of the conferences of the parties are generally small; they seldom control significant resources that could make them hotbeds of corruption. Meetings of conferences of the parties are essentially diplomatic events. Participants acting as agents of member states may be more or less effective in promoting the interests of their principals. But there is little room for corruption in the ordinary sense in such settings. Collusion among the representatives of two or more states to achieve some outcome antithetical to the interests of a regime's other members, for instance, is a fairly remote possibility. This suggests that accountability at the international level is not a critical concern when it comes to the effectiveness of international regimes.

Much of the day-to-day work of administering the provisions of international regimes ordinarily falls to personnel associated with agencies of the governments of members states assigned to handle the relevant tasks. In the United States, for example, the Environmental Protection Agency is the lead agency charged with implementing the provisions of the Montreal Protocol regarding ozone-depleting substances; the National Oceanographic and Atmospheric Administration plays this role regarding the provisions of the whaling regime, and the Fish and Wildlife Service located within the Department of the Interior is the lead with regard to international agreements dealing with threats to terrestrial species. Issues of accountability regarding the actions of agencies located within the governments of member states may become a matter of serious concern. What is notable in this regard, however, is that lead agencies frequently become advocates for the regimes entrusted to them within the broader political processes of the governments within which they operate (Dai 2007). The EPA can be counted on to fight hard for compliance with the terms of the Montreal Protocol; NOAA and the FWS will do the same with regard to regimes for threatened or endangered species for which they are responsible. And there is nothing unusual about the United States in this regard. Under the circumstances, accountability may not be a critical issue when it comes to the effectiveness of international regimes.[11] But larger economic and political forces that sap the capacity of governments to carry out their obligations or marginalize the issues that individual regimes address can play a determinative role in undermining the effectiveness of specific regimes. Here, too, a question regarding scope conditions comes into focus. What can we say about the broader forces that determine whether those in charge of carrying out the terms of specific regimes are willing and able to perform their roles in an effective manner?

What are the next steps for those interested in understanding the links between good governance and effective governance? Needless to say, the development of an operational metric for assessing the effectiveness of regimes and tracking trends in effectiveness over time would be a big help. This would open up opportunities to formulate and test hypotheses about the importance of such things as specific types of participation (e.g., different decision rules governing the work of conferences of the parties) or various approaches to transparency (e.g., particular MRV procedures). But I am not optimistic about the prospects for significant progress in this realm. My colleagues and I made a concerted effort to measure effectiveness in creating the International Regimes Database during the 1990s and early 2000s (Breitmeier, Young, and Zürn 2006). We did make some progress in this effort; some of the findings regarding the relevance of consensus practices in the decision-making of regimes are certainly relevant to the focus of this chapter. But the effort that went into developing this database made it clear to me at least that there are major problems with the construction of precise and useful metrics in this realm. No one has made the effort to extend the project's assessment of effectiveness beyond the original cutoff date of 1999, and no one has developed an alternative approach to the measurement of effectiveness that does a better job of capturing effectiveness empirically or that is more tractable (Mitchell 2008; Underdal 2008).[12] Research on international regimes has moved on to address other topics (e.g., institutional interplay, the role of nonstate actors, the significance of behavioral mechanisms that are hard to represent in utilitarian terms). There is no indication that a new and more tractable metric for measuring effectiveness will become available anytime soon.

On the other hand, additional research on the conditions under which the PTA variables make a difference regarding the effectiveness of international regimes strikes me as a promising line of inquiry. Are there situations in which regulatory tasks can be handed off to nonstate actors, such as the International Association of Classification Societies in the case of certifying that new ships meet standards established by the International Maritime Organization or the Marine Stewardship Council in the case of certifying that specific fisheries are conducted in a sustainable manner? Developments of this sort might alter our thinking about the need for participation on the part of states in the performance of some regulatory tasks in international society. Are there ways to make use of advanced technologies to confirm compliance with the provisions of international regimes or to identify cases of efforts to violate important provisions clandestinely? There is no doubt that the advent of satellites capable of performing inspections without running into the sensitivities associated with on-site inspection played a role of considerable importance with regard to the development and administration of arms control agreements dealing with strategic weapons. It is possible that similar technological developments will play a key role in addressing problems of MRV in other issue domains. Automatic Identification Systems, for example, can support efforts to confirm compliance with safety regulations for commercial vessels and to identify violators of regulations regarding restricted areas in marine fisheries.

Are there ways to ensure that lead agencies in the governments of member states charged with carrying out the terms of international regimes are sheltered effectively from the impacts of contending political forces? The development of sizable trust funds not subject to routine budgetary battles, for example, might go some way toward addressing this problem. Interestingly, there is considerable experience with the creation and operation of trust funds at the international level (Sand 1994). What might well emerge from research relating to questions of this sort could be a clearer sense of the conditions under which the PTA variables make a difference to the pursuit of effectiveness in the establishment and operation of international regimes.

Concluding Observations

Does good governance produce effective governance, especially with regard to international regimes pertinent to the pursuit of sustainability in the Anthropocene? The phrase good governance is susceptible to a wide range of interpretations. In this chapter, I have taken it to refer to a set of process considerations rather than to matters relating to outcomes or impacts, like the extent to which governance systems produce results that conform to our views regarding fairness or equity. Thus, I have directed attention to what I call the PTA variables—participation, transparency, and accountability—and asked whether we can show that these factors are associated with effectiveness particularly at the level of problem-solving.

The bad news is that we have no simple or clear answer to the question I have posed. This is partly a matter of severe limitations in measuring effectiveness, much less in developing an index that would allow us to rank or track regimes in some sensible way in terms of their success in solving problems. In part, it arises from the fact that analysts have not made a sustained effort to assess the role of the PTA variables in their studies of regime effectiveness. But there is also some good news to report regarding efforts to understand the significance of good governance. Given the nature of international society, a high level of participation on the part of member states may be a necessary condition for regime success. A high level of transparency, on the other hand, may be more important regarding regimes operating in some issue domains than in others. In my judgment, looking more closely at the mechanisms involved and the conditions affecting the extent to which the PTA variables play important roles as determinants of the success of regimes is likely to constitute the cutting edge in thinking about the relationship between good governance and effective governance as we move deeper into the Anthropocene.

Conclusion: Science, Policy, and the Governance of Complex Systems

Introduction

If the thesis I have presented in the preceding chapters is persuasive, then we must set about overcoming the grip of business as usual in our thinking about the governance of complex systems, and guarding against the power of path dependence in our efforts to create and implement governance systems that prove effective in the pursuit of sustainability in the Anthropocene. This is not a call for abandoning ongoing research programs and discarding the existing repertoire of policy prescriptions for solving problems involving human-environment interactions. But there is an obvious need to upgrade the social capital available to those who are responsible for coming to terms with this challenge.

In this concluding chapter, I address this need in three steps. The first substantive section focuses on research opportunities, identifying promising directions for members of the science community who seek to add to the pool of knowledge available to those responsible for meeting the challenge of governing complex systems. The second section turns to the realm of policy and policymaking; it identifies pitfalls hampering current efforts to cope with the problems of the Anthropocene, and highlights new directions that can enhance social capital and strengthen our capacity to address these issues more effectively in the future. The final section then turns to the science/policy interface in the light of the preceding discussion. It makes the case for active engagement between members of the research community and members of the policy community who stand to benefit from each other's contributions to the pursuit of sustainability in a world of complex systems. While there are legitimate concerns in this realm about the dangers of co-optation undermining the force of social criticism, there are compelling reasons to encourage the melding of analysis and praxis on the part of those who share a commitment to advancing the cause of sustainability in the Anthropocene.

Research Opportunities

Research on governance and governance systems has grown rapidly over the course of recent decades. Much of the innovation in this field stems from the work of those who focus on the role of social institutions in addressing needs for governance to come to terms with large-scale environmental problems, ranging from long-range transboundary air pollution and the thinning of the stratospheric ozone layer, to the growth of dead zones in the oceans and the destruction of tropical forests in terrestrial systems (Rosenau and Czempiel 1992; Ba and Hoffmann 2005; Biermann et al. 2009; Young 2013b). We have learned a lot about social institutions construed as problem-solving mechanisms, and about how these institutions interact with a variety of biophysical and socio-economic forces to determine the trajectory of human-environment relations. But now we need to focus on the specific challenges of governing complex systems under the conditions prevailing in the Anthropocene. In this section, I take up this subject, discussing thematic issues that are ripe for analysis in this context and directing attention to the methodological hurdles facing those who seek to make progress in addressing these issues.

Transcending the Regulatory Paradigm

Our thinking about governance is structured to a striking degree by what we may describe as the regulatory paradigm. This paradigm approaches efforts to meet needs for governance as a matter of formulating rules spelling out behavioral prohibitions and requirements, promulgating regulations to operationalize these rules with regard to concrete situations, developing mechanisms to elicit compliance from identifiable subjects, making use of sanctions to ensure that compliance is forthcoming, and providing authoritative procedures for settling disputes regarding the interpretation of the rules or their application to concrete cases. Underpinning this mode of thought is the belief that problem-solving is fundamentally a matter of advancing the influence of the rule of law through a process that leading thinkers describe as "making law work" (Zaelke et al. 2005). As the title of a prominent text in the field puts it, the core concern is a matter of achieving "compliance with international regulatory agreements" (Chayes and Chayes 1995).

This way of thinking clearly has merit; there is much to be said for ongoing efforts to strengthen the rule of law in international society. But the regulatory paradigm is not the only lens we can employ in thinking about the nature of governance, and about the options available to those seeking to make progress toward solving problems in settings that exhibit the features of complex systems. In fact, the regulatory paradigm, with its emphasis on entrenching well-defined rules and creating formal systems to ensure compliance on the part of subjects, may impede problem-solving in situations featuring transformative changes that put a premium on an ability to adjust nimbly or agilely to changing circumstances. It follows, in my judgment, that meeting the challenge of governing complex systems calls for exploring alternative ways to think about problem-solving in the interests of supplementing, though not necessarily replacing, the regulatory paradigm as we endeavor to augment our toolkit for achieving sustainability in the Anthropocene.

As it turns out, there is no lack of alternatives to the regulatory paradigm that may prove helpful in coming to terms with the challenge of governing complex systems. One interesting option centers on what I characterized in chapter 6 as principled governance. Principles are not rules. They are guidelines that are helpful in weighing the pros and cons of alternative courses of action, rather than prohibitions or requirements that call for compliance regardless of the specific circumstances at hand. Principles provide normative guidance, but they are situational in the sense that their application requires a consideration of the relevant features of specific cases. The principle of common but differentiated responsibilities, for example, calls for sensitivity to differences in the circumstances of individual actors, and for calibrating expectations regarding the contributions of specific actors to solving particular problems according to rough calculations of their capacity. It does not impose a rigid formula that all parties concerned are expected to comply with as a matter of course. In the case of climate change, for example, this principle calls for developed countries to assume a larger share of the burden of reducing emissions of greenhouse gases than developing countries assume. But it does not prescribe quantified targets for inclusion in the

Intended Nationally Determined Contributions (INDCs) of individual countries or categories of countries. Needless to say, principled governance has drawbacks. Individual actors may choose to ignore relevant principles, or adopt self-serving analyses of prevailing situations to justify shirking what most observers would regard as their real responsibilities. Nevertheless, principled governance seeks to make progress in steering social groups by drawing the individual members of the group into a web of normatively based expectations, rather than articulating prescriptive rules focusing attention on issues of compliance and the use of enforcement mechanisms to strengthen the incentives of individual members to comply with the rules.

Another lens that differs more fundamentally from the regulatory paradigm arises in connection with what I described in chapter 5 as governance through goal-setting in contrast to governance through rule-making. Goal-setting is a matter of establishing well-defined priorities, allocating resources to match the priorities, and then encouraging the members of the relevant group to join forces in a campaign designed to achieve agreed upon goals. The case of the Millennium Development Goals (MDGs) is instructive in this regard. No one assumed legally binding obligations to make specified contributions toward the fulfillment of the MDGs. Yet, there is general agreement that the MDGs have served to enhance the prominence of key goals (e.g., eliminating extreme poverty, eradicating some diseases) and to galvanize efforts to make a collective push toward fulfilling some of these goals (Sachs 2015). It would be a mistake to exaggerate the effectiveness of goal-setting as a governance strategy. There is an understandable but unjustified tendency to ascribe causal force to the MDGs that goes well beyond what any rigorous examination of the evidence will support. And as the recent experience with reaching agreement on the Sustainable Development Goals (SDGs) makes clear, it is hard to avoid the trap of agreeing to such a large number of goals that it is difficult to justify treating any one of them as a true societal priority. Still, the logic of goal-setting as a governance strategy involves behavioral premises that differ fundamentally from those embedded in rule-making as a governance strategy (Kanie and Biermann 2017). Among other things, goal-setting may offer a more flexible approach to governance than rule-making with its emphasis on entrenching rules and developing the apparatus needed to enforce them uniformly and indefinitely.

Nor does this exhaust the range of options worthy of consideration in thinking about the governance of complex systems. In this book, I have focused for the most part on state-centric governance strategies. But there is growing interest in the role of a variety of nonstate actors in addressing needs for governance. In some cases, as in accounts of corporate social responsibility, analysts have explored circumstances under which private actors are willing and able to tackle problems in the absence of intervention on the part of the state (Lyon and Maxwell 2004; Delmas and Young 2009). In other cases, as in strategies featuring public–private partnerships, the focus is on opportunities for states and nonstate actors to join forces to address problems that neither side can solve on its own. It is important to avoid naïve conclusions about the effectiveness of these approaches to dealing with needs for governance. Strategies emphasizing the role of corporate social responsibility, for example, are unlikely to prove effective in the absence of well-defined and widely accepted institutional arrangements dealing with property rights, contracts, liability rules, and dispute settlement (Reich 2015). Nevertheless, there is every reason to explore these options systematically in the interests of expanding the range of strategies available to deal with the governance of complex systems.

The take-home message is clear. There is much to be said for casting a wide net as we search for ways to meet needs for governance that are well suited to the challenges of governing complex systems. Once we break the bonds of path dependence, additional approaches are likely to come into focus. We should be alert to new ways to think about transcending the regulatory paradigm, and open to adding new governance strategies to the palette of tools available to those seeking to achieve sustainability in the Anthropocene. The goal is not to replace one dominant paradigm with a single alternative. Rather, success in meeting the challenge of governing complex systems is likely to require the development of a portfolio of governance strategies, and the cultivation of skills in diagnosing needs for governance arising in specific situations and in selecting suitable governance strategies to solve specific problems.

Linking Institutions and Organizations

It is no secret that creating institutional arrangements without providing adequate capacity to implement them effectively is a recipe for failure.

The phrase "unfunded mandates" has become well known as shorthand for this proposition. This is a serious concern at all levels of social organization. Even in mainstream settings, there is a pronounced tendency to devise rules and regulations without assigning responsibility to specific public agencies for administering them, or providing lead agencies with the resources needed both to administer them on a day-to-day basis and to elicit compliance on the part of subjects. But the absence of anything resembling a government in the conventional sense in international society makes this concern especially acute in efforts to address large-scale and transboundary problems. Any serious examination of the raft of modern multilateral environmental agreements makes it clear that most of these governance systems have little capacity of their own to deal with matters of implementation and compliance, a fact that makes them dependent on the ability and the will of member states to implement the provisions of these agreements within their own jurisdictions.

Less familiar but equally troubling is the opposite problem: the creation of organizations in the absence of well-constructed institutional arrangements or governance systems for them to administer (Young 2008a). Situations of this kind give rise to costly and often oppressive bureaucracies that lack a well-defined rationale, and that consequently precipitate the complaints heard over and over again about the costliness of big government and the frustrations associated with the proliferation of bureaucratic red tape. In one sense, this problem is less severe at the international level than it is in domestic settings, simply because government in the conventional sense is so limited in international society. But as critics of the UN System point out tirelessly, this does not mean that there is no problem at the international level of agencies that are long on bureaucracy but short on well-defined governance arrangements to administer, and that are more concerned with protecting their turf than with forming the coalitions needed to address complex problems. Given the costs—calculated in terms of bureaucratic oppressiveness as well as in conventional monetary terms—of excessive bureaucracy, the burden of proof should always fall on those who propose to create new administrative agencies.

From the point of view of governing complex systems, however, the critical issue to explore is the need to match institutions and organizations. If we are operating within the framework provided by the regulatory paradigm, the emphasis is likely to fall on building organizational capacity to handle matters of monitoring, reporting, and verification (MRV in the terminology of the climate regime) and to administer the sanctions needed to enforce compliance with the rules of the game. But this sort of capacity is both costly in monetary terms and politically sensitive. Subjects who interact with administrative arrangements of this sort are apt to experience a sense of resentment at the intrusiveness of MRV procedures as well as a sense of being coerced by those responsible for the operation of enforcement mechanisms. And situations of this sort may well spark an action/reaction process between subjects and administrators that fails to produce winners and losers but that is inevitably costly to society.

Clearly, there is a need for organizational arrangements of this sort under some conditions. There is no getting around the need to monitor compliance with rules and to operate enforcement mechanisms to maximize compliance in some situations. But this is a costly way of governing complex systems in the interests of achieving sustainability in the Anthropocene. What is needed in this setting are organizations that can play a role in redefining how we conceptualize problems and in "nudging" actors to place appropriate values on ecosystem services, to adopt social rates of discount, and to develop perspectives on well-being that are not limited to material gains (Thaler and Sunstein 2008). The burgeoning literature on what has become known as behavioral economics offers an array of insights in this regard that may well prove helpful to those responsible for governing complex systems (Akerlof and Shiller 2011; Kahneman 2011; Thaler 2015). At a minimum, we should be exploring these lines of inquiry rather than simply struggling to formalize rules and regulations and to build more effective enforcement systems to elicit compliance from subjects.

Combining Durability with Nimbleness or Agility

A central challenge going forward in efforts to govern complex systems is to find ways to achieve nimbleness or agility in responding to nonlinear changes, abrupt transitions, and surprising developments without sacrificing the durability of governance systems needed to make subjects take them seriously as guides to behavior. A governance system whose key provisions are subject to continual change will not be effective in solving problems. The reason is simple. Subjects who perceive that key provisions are ephemeral will have no incentive to adjust their behavior to comply with prohibitions and requirements, to take principles seriously, or to engage vigorously in the collective efforts needed to fulfill common goals or objectives. In efforts to govern complex systems, however, durability can become part of the problem rather than part of the solution. We need governance systems that can adapt easily to changing circumstances, and that are nimble or agile in responding appropriately to surprises or to challenging circumstances that are unanticipated. The now familiar concept of resilience is a partial response to this concern (Gunderson and Holling 2002; Folke 2006; Walker and Salt 2006). But resilience does not provide a roadmap that can help those responsible for meeting this challenge of balancing needs for nimbleness or agility and durability.

What is needed in terms of research is the identification and elaboration of strategies that can strike a balance between nimbleness or agility and durability under real-world conditions. There are some leads that are worth examining carefully in this realm. One idea arises from the distinction between constitutive institutions and operational regimes. It may make sense to differentiate between the two types of arrangements, investing constitutive provisions with a relatively high degree of durability, while allowing operational regimes to adjust more quickly and easily to changing circumstances. Similarly, there is a case for introducing institutional thresholds allowing for a shift in the balance between durability and nimbleness or agility once a well-defined threshold is passed. This is the idea underlying the declaration of a state of emergency granting public officials the authority to bypass certain normal procedures so long as the emergency lasts. It is easy to find examples involving the granting of authority to officials during times of war that would be incompatible with the principles of democracy during normal times.

We need to explore analogs to such situations relevant to the pursuit of sustainability in the Anthropocene. This is an area calling for intense efforts to come up with suitable institutional innovations. As the case of climate change makes clear, we face a dilemma caused by the mismatch between the dynamic character of the problem and the sluggishness of our institutional processes for coming to terms with it (Young 2010). The agonizingly slow process of negotiating the terms of a new legally binding agreement founded on the platform provided by the UN Convention on Climate Change contrasts sharply with the pace of change in the Earth's climate system. The dramatic failure to reach consensus on the terms of an agreement in Copenhagen in 2009 produced the shift toward voluntary pledges known as INDCs during the run-up to the meeting of the Conference of the Parties in Paris in 2015. But even this shift has failed to produce pledges that come close to fulfilling the requirements for reductions in greenhouse gas emissions needed to meet the goal of limiting increases in temperature at the Earth's surface to no more than 2°C, much less to no more than 1.5°C. The Paris Agreement offers some basis for hope that participants in the climate regime will devise an effective method for strengthening or ratcheting up their commitments at regular intervals (Paris Agreement 2015), and it may be that severe climate shocks during the near future will serve to break the grip of path dependence with regard to climate governance. But this case presents a dramatic illustration of the problem of striking a suitable balance between durability and nimbleness or agility in efforts to govern complex systems.

Acting in the Face of Uncertainty

There is no escaping the need to make decisions under conditions of uncertainty as we endeavor to govern complex systems in the Anthropocene. The problem is not just a matter of attaching well-grounded probability estimates to a range of possible outcomes. We do not know how close we are to the thresholds associated with various planetary boundaries (Rockström et al. 2009; Steffen et al. 2015). We have little ability to calculate the probability of the occurrence of trigger mechanisms associated with key thresholds or tipping points (Lenton et al. 2008). We are poorly equipped to recognize the difference between the cyclical dynamics of oscillations and the directional forces producing bifurcations. Our estimates of the costs to society of crossing such thresholds and moving outside the "safe operating space for humanity" are highly subjective at best. Yet we cannot avoid making decisions in situations of this type, especially once we realize that doing nothing is in effect a decision to continue with business as usual. It follows that we face a situation that cannot be solved simply by calling for more and better research to eliminate, or at least to alleviate, the problem. We must come to grips with the necessity of making choices among options characterized by high levels of uncertainty regarding the nature of the options available as well as the results of pursuing one option rather than another.

Yet, this does not mean that we must choose blindly in situations of this sort. While much more research is needed regarding decision-making under uncertainty, a few initial insights are worth noting at this stage. To begin with, the precautionary principle, which suggests that it is desirable to hold off on action until more is known about the situation, is a sensible response to uncertainty in some situations but not in others. It may make sense, for example, to impose a moratorium on harvesting a particular stock of fish until we know more about the population dynamics of the relevant species. But this is not necessarily good advice when it comes to addressing the issue of reducing emissions of greenhouse gases. The potential costs associated with what are known as rapid climate change events (RCCEs), for example, are such that it may make more sense to invest substantial resources in minimizing the likelihood of such events occurring, even if the investment turns out to be unnecessary, than to avoid such investments under conditions where they turn out to be essential (Alley 2000; Mayewski and White 2002). Under such conditions, a reasonable strategy is to proceed in a stepwise fashion, dividing actions aimed at addressing problems into a number of stages or phases and taking stock of the results following each stage in the process. Procedures of this sort can combine the resoluteness required to come to terms with serious problems with the flexibility needed to make mid-course corrections in the implementation of decisions based on monitoring the results on a continuous basis.

There is much to be said as well for identifying biases that regularly impact decision-making under uncertainty and minimizing the impact of these biases on actual choices. There is a pronounced tendency, for example, to exaggerate the expected costs of addressing large-scale environmental problems in contrast to the costs associated with failure to address them. Repeatedly, we have found that the actual costs of addressing problems (e.g., acid precipitation, stratospheric ozone depletion) are a fraction of the costs projected by various participants in the debate about whether or not to take concrete steps to address the problems. In many cases, this is attributable to the concerns of special interests (e.g., coal companies worried about losing markets for their product) rather than to realistic estimates of the relevant costs. In other cases, it stems from the emergence of innovative responses, once relevant actors acknowledge the necessity of coming to terms with the problem and redirect their energy from opposing actions to searching for cost-effective solutions (Parson 2003). A corollary to this argument involves the value of no-regrets policies in which outcomes produce societal benefits regardless of their effectiveness in addressing the initial problem. Reductions in greenhouse gas emissions offer clear-cut examples of this proposition. Reducing the combustion of fossil fuels and especially coal is likely to produce unambiguous improvements in public health, for example, whether or not they are needed to avoid transgressing the planetary boundary relating to climate change. The evidence now linking heart disease and strokes to the density of airborne pollutants (e.g., PM 2.5), for instance, is sufficiently convincing to justify actions to reduce the consumption of coal in a country like China, quite apart from the role that burning coal plays with regard to the problem of climate change.

Addressing the Methodological Challenge

Before concluding this section, I want to make some observations about methodological hurdles facing those seeking to respond to the research opportunities I have identified. The core methodological challenge for those conducting research on governance systems centers on the role of complex causality. In real-world situations, many factors interact to determine the trajectories of socioecological systems, a phenomenon that is intensified in the case of complex systems featuring teleconnections, cascades of change, and emergent properties. The challenge then is to identify the signal of social institutions, pinpointing the roles that institutional arrangements play in such settings, and revealing the ways in which institutions interact with both socioeconomic forces and biophysical drivers to determine observed outcomes. If anything, this challenge is intensified by the fact that we are ordinarily dealing with causal clusters rather than causal chains, so that it does not make sense to focus attention on individual links in the chain in the interests of simplifying the analysis of causal connections (Young 2002a).

The most common response to this challenge is to emphasize the danger of false positives or, in other words, conclusions that overestimate the significance of institutions in explaining observed outcomes. In some respects, this is an expression of healthy skepticism. Various actors, ranging from political leaders to a variety of nongovernmental organizations, tend to exaggerate the effects of governance systems, especially in cases where they are able to make reasonable claims to being responsible for the creation or implementation of these systems. It is not surprising, therefore, that there are numerous contributions to the scientific literature on international environmental regimes arguing that the claims of those who believe that specific regimes (e.g., the long-range transboundary air pollution regime in Europe) have proven to be effective are unproven at best and may involve relationships that are largely spurious.

In my judgment, however, there is an equally pressing need to be concerned about the danger of producing false negatives. It is all too easy to conclude that outcomes are driven by the distribution of political power, the uncontrollable influence of technological forces, or the biophysical components of the Earth system, so that social institutions are merely epiphenomena whose impacts on the trajectory of socioecological systems are more apparent than real (Strange 1983). Among other things, arguments of this sort are congenial to those who are beneficiaries of existing arrangements (e.g., the oil companies in the case of climate change) and who therefore wish to suppress efforts to introduce changes in prevailing governance systems. What is needed in this case is a recognition that just because institutional arrangements are only elements in causal clusters that collectively determine outcomes does not mean they are insignificant parts of the equation.

The way forward in my judgment is to adopt a portfolio approach to research methods and a focus on conditions under which institutions make a difference (i.e., scope conditions), rather than to join the debate regarding whether or not institutions are worthy of attention on the part of those seeking to understand the behavior of complex systems (Young et al. 2006). Findings that are corroborated by quantitative and qualitative modes of analysis as well as by other procedures (e.g., Qualitative Comparative Analysis, meta-analysis) carry a lot more weight than findings derived from a single mode of analysis. An important outcome of adopting this methodological strategy is likely to be progress toward the identification of conditions under which various governance systems make a difference. The same governance system that works relatively well in dealing with one problem may fail miserably in addressing other problems. This is particularly true of what are often described as policy instruments, such as tax policies, cap-and-trade mechanisms, permit systems, and so forth. Conversely, we may find that there is more than one institutional solution to specific problems. Both planning systems and market-based regulatory systems, for example, may produce the desired results in efforts to come to terms with problems like the emission of greenhouse gases (Young et al. 2015). While progress in identifying conditions determining the effectiveness of governance systems may well be slow, an approach of this sort seems essential in developing the research agenda for the next phases of research on governing complex systems.

Lessons for Policy-Makers

What lessons should policy-makers draw from my analysis of the issues involved in governing complex systems? Here, too, the proper inference is not that we need to abandon existing practices and make a decisive switch to a new mode of operation. Rather, we need to expand our toolkit and think in diagnostic terms in order to tackle the problem of fit in specific situations. We also need to take advantage of processes that can alert us to the unintended consequences of policy options that are appealing in superficial terms. In this section, I elaborate on this theme, challenging some prominent features of business as usual, sketching some additions to the existing repertoire of practices, commenting on the uses of assessments, simulations, and scenario exercises, and illustrating the contribution of new technologies to coming to grips with a variety of problems arising in human-environment relations.

Rethinking Orthodox Preferences

There is a distinct preference in policy circles for embedding the provisions of governance systems in legally binding agreements supported by the creation of "normal" intergovernmental organizations. The gold standard is a regime whose provisions are set forth in an international convention or treaty and that is administered by an intergovernmental organization that has legal personality and access to funds provided under the terms of an established indicative budget. The World Trade Organization (WTO) is a familiar example. The assumptions underpinning this preference are easy to identify. Legally binding agreements are expected to generate a normative pull increasing pressure on the parties to comply with their provisions. Normal intergovernmental organizations are needed to ensure that such agreements are administered properly and that adequate resources are available to deal with issues of compliance and enforcement.

No doubt, these assumptions have merit under some conditions. But, as the case of climate change demonstrates clearly, they do not provide a universal recipe for success in governing complex systems. Legally binding agreements are difficult to negotiate, and key parties may fail to ratify them even when their provisions are watered down to avoid political sensitivities. Above all, such arrangements are sticky in the sense that it is the political equivalent of pulling teeth to revise or restructure them to address changing needs. The effort to restructure the 1997 Kyoto Protocol to deal effectively with the growing challenge of climate change, for example, has taken years; the result is a new agreement whose legal character is ambiguous, whose major provisions amount to voluntary pledges, and that is not expected to become fully operational before 2020. Meanwhile, the concentration of greenhouse gases in the Earth's atmosphere continues to rise to levels that have already produced temperature increases of ~1°C and that may well make it impossible to meet the widely accepted goal of preventing increases in temperatures at the Earth's surface in excess of 2°C, much less the emerging preference for 1.5°C (Intergovernmental Panel on Climate Change 2014; Paris Agreement 2015).[1]

What is more, success in addressing some problems does not require arrangements that meet the expectations of the gold standard. Regimes that focus on generative tasks (i.e., identifying emerging issues, framing them for consideration in policy forums, devising a discourse in which to embed them, and pushing them to the forefront of public agendas), for instance, do not depend on legally binding provisions to make a difference and need not require the services of a normal intergovernmental organization. The Arctic Council, a high-level but informal and lightly administered body, has achieved considerable success in these terms (Young 2011b). Similarly, regimes that are able to build effective bridges to nonstate actors may succeed without the support of elaborate organizational capacity. The International Convention for the Prevention of Pollution from Ships (MARPOL), for example, has tackled the problem of reducing oil spills from tankers through a system that creates key roles for the International Association of Classification Societies (IACS), a nongovernmental organization, and the marine insurance industry, a collection of private actors (Mitchell 1994). The take-home message here is that we should make a concerted effort to design institutional arrangements to address needs for governance in a manner that is tailored to the key features of the relevant issue areas, that takes advantage of opportunities to find roles for existing players, and that minimizes the need to build intergovernmental organizations that are costly to operate and likely to fall prey to a variety of bureaucratic problems.

A particularly important concern in this realm is the tendency to confuse the roles of institutions and organizations and, as a result, to endeavor to solve institutional problems through the creation of new or restructured organizations (Young 1989). Institutions are systems of rights, rules, principles, norms, and social practices intended to govern the behavior of those who are active in a given issue area. Organizations, by contrast, are material entities that have offices, personnel, budgets, and legal personality. The UN Charter, for example, lays out a system of institutional arrangements; the UN Organization is a material entity created to administer the provisions of the charter. In most cases, the implementation and operation of institutions requires organizational capacity. But organizations not designed to fit the needs of specific institutional arrangements seldom produce positive results; poorly designed organizations can easily turn into sluggish bureaucracies that lack well-defined rationales and can become part of the problem rather than part of the solution.

This is why many efforts to reform international organizations turn into exercises in futility. Most proposals to turn the UN Environment Programme into a proper organization (a UN Environment Organization or even a World Environment Organization), for example, suffer from this defect (Biermann and Bauer 2005). There is no reason to expect a UNEO or a WEO to achieve great success so long as it is not tied to a restructured system of international environmental governance. The contrast with the WTO is instructive in this regard. Although it is now under pressure from a number of sources, most observers regard the WTO as a successful organization. But it is essential to note that much of this success is attributable to the fact that the WTO administers a complex regime whose provisions constitute an integrated system of institutional arrangements governing trade between or among countries that are accepted by all the major players in international society. Should shifting circumstances undermine the acceptability and the legitimacy of the trade regime, there is little that the WTO would be able to do to stem the growing tide of pressures for reform and restructuring.

Similar observations are in order with regard to other organizational arrangements (Kanie et al. 2012; Biermann 2014). There is general agreement that the UN Commission on Sustainable Development, created as a product of the 1992 UN Conference on Environment and Development, was a failure. As a result, the participants in the 2012 Rio+20 conference, agreed to dismantle the CSD and to replace it with a body to be known as the UN High-Level Political Forum on Sustainable Development (HLPF) (Bernstein 2013). This forum, which took over from the CSD in 2013, has a number of distinctive features. It is meant to have a dynamic and action-oriented agenda, and it will meet every fourth year at the Heads of Government or State level. These are progressive reforms. In my judgment, however, the prognosis for the HLPF is not particularly good. It represents an organizational fix to problems that are essentially institutional in nature. There is nothing in this transition, for instance, that will strengthen materially the climate regime, the regime for biodiversity, or the regime to combat desertification as governance systems capable of steering the actions causing the relevant problems in an effective manner. None of this is to say that organizational capacity is unimportant or that we should not devote high-level attention to the design of appropriate organizations. But it is to say that in seeking to govern complex systems we need to beware of initiatives that amount to futile efforts to make up for institutional shortcomings through organizational reforms.

Expanding the Repertoire of Practices

Policy-makers concerned with the development of governance systems display a strong tendency to reason by analogy, applying solutions that appeared to be successful in the past to the framing of arrangements designed to address new problems. Up to a point, this is a perfectly reasonable mode of operation. But the challenge of governing complex systems is all about differences between the problems of the Anthropocene and the more familiar problems of the past. What this means is that there is a need to expand the toolkit available to those responsible for the development of regimes to govern complex systems. In this subsection, I comment on this need from the perspective of (i) mainstream intergovernmental practices and (ii) innovative practices that feature enhanced roles for a variety of nonstate actors.

With regard to the first of these perspectives, the essential challenge is to maintain flexibility without sacrificing the behavioral force of the institutional arrangements we adopt. In fact, we are already experimenting with a variety of innovations designed to meet this need. The ozone regime, widely regarded as the most successful of the modern international environmental agreements, offers a number of examples (Parson 2003). Amendments calling for accelerations of the phaseout schedules for ozone deleting substances regulated under the terms of this governance system do not require ratification to take effect. The Montreal Protocol establishes a non-compliance procedure that emphasizes the importance of working with members to solve problems of compliance rather than subjecting them to penalties or threats of penalties to coerce compliance with the regime's requirements. The Montreal Protocol Multilateral Fund, established as a self-contained body, has achieved an excellent track record in using relatively modest funds to help developing countries move forward in ways that avoid dependence on the use of ozone-depleting substances. And along the way, this regime has presided over changes resulting in more reductions in emissions of greenhouse gases than the UN Framework Convention on Climate Change (Velders et al. 2007). In this regard, the current debate over proposals to phase out the use of hydrofluorocarbons (HFCs) is particularly significant (Zaelke et al. 2016). Used in a number of applications as a substitute for ozone-depleting substances, HFCs turn out to be powerful greenhouse gases even though they are not ozone-depleting substances. A concerted effort is currently making headway on plans to phase out HFCs under the terms of the Montreal Protocol as an important contribution to the effort to address the problem of climate change.

Although it is a particularly striking case, the ozone regime is not the only governance system that takes advantage of flexibility mechanisms in the search for solutions to significant problems. Consider the Polar Code applying to ships operating in Arctic and Antarctic waters, adopted by the International Maritime Organization in 2015, as another source of illustrations (VanderZwaag 2012). The Polar Code will enter into force, most likely at the beginning of 2017, in the form of amendments to the Safety of Life at Sea Convention (SOLAS) and MARPOL, because amendments to these conventions enter into force automatically if no country lodges an objection within a specified period of time. Similarly, the code deals with a number of issues of compliance by shifting the burden of proof to vessel owners/operators. Vessels subject to the code are required to carry valid and up-to-date Polar Certificates. Automatic Identification Systems (AISs) are able to identify ships operating in polar waters using satellite observations and then to access the relevant database electronically to determine whether a specific vessel has an up-to-date Polar Certificate. The cost of operating the necessary AISs is modest, and this procedure puts the burden of proof on the owner/operator of a vessel to demonstrate compliance. Any vessel that is identified publicly as non-compliant is unlikely to win favor either with shippers or with insurers.

As the shipping example suggests, we also need to take into account the shifting character of international society and to include important nonstate actors in thinking about ways to govern complex systems effectively. IACS, for example, is a powerful player in maritime affairs. It has the authority to accredit new ships or, in other words, to certify that they meet all standards applicable at the time they enter into service. Without a valid certificate, a ship is unlikely to be able to obtain insurance. And there is little market for ships that are uncertified and uninsurable (Mitchell 1994). So, while the standards are established under the provisions of intergovernmental agreements, nonstate actors are powerful players with respect to implementation, a fact of considerable importance given that IMO does not have the capacity itself to play these roles effectively. Continuing with the same example, IACS operates a system of ice classification for ships operating in polar waters. This system encompasses seven polar classes to be used in determining whether ships are constructed and operated in a manner that allows them to operate safely under a range of Arctic and Antarctic conditions. Here, too, a working relationship with IMO and the insurance industry makes this system effective in providing guidance for those in charge of operating individual ships under more or less severe conditions.

Some analysts go further, arguing that corporate social responsibility (CSR) is a powerful force that we can rely on to address a range of concerns relating to sustainability in the Anthropocene (Lyon and Maxwell 2004; Vogel 2006; Delmas and Young 2009). Without doubt, CSR is a significant phenomenon. In some cases, it is simply a matter of good business practices to improve efficiency in a way that reduces environmental impacts or takes advantage of new environmentally benign technologies to develop profitable products. Yet there is no basis for concluding that we can count on CSR to solve large-scale environmental problems, thereby eliminating the need to create and administer governance systems in the ordinary sense of the term. Much more likely is a situation featuring complex relations between those responsible for administering governance systems and those in charge of corporations and other actors subject to the provisions of these systems. Even so, this is an important point. It suggests that the pursuit of sustainability in the Anthropocene may be better regarded as a search for mutually acceptable arrangements, than as a matter of enforcing regulations aimed at actors that are always on the lookout for ways to shirk their obligations or to launch initiatives designed to test the limits of what they can get away with without incurring severe sanctions.

Making Use of Decision Support Tools

It is necessary to make decisions in the face of uncertainty in addressing needs for governance arising in every setting. Ironically, however, the onset of the Anthropocene has had the effect of increasing levels of uncertainty regarding many matters of governance. Despite rapid advances in knowledge and massive increases in our capacity to monitor relevant variables, the dynamics of complex systems are far more difficult to understand than those of the more familiar systems of the past. Yet this does not mean that policy-makers are helpless in the face of uncertainty, reduced to making blind choices or following the dictates of non-falsifiable ideologies when it comes to devising governance systems. Among the tools available to assist those who must make policy choices under uncertainty are assessments, simulations, and scenarios. Taken together, these procedures are sometimes described as decision-support tools.

Assessments take the form of sustained efforts to assemble and evaluate the state of knowledge regarding systems of particular interest to policy-makers, providing some indication of confidence levels regarding propositions about the behavior of these systems, and highlighting the major gaps in our current understanding. Perhaps the most prominent examples are the periodic assessment reports on climate prepared by the Intergovernmental Panel on Climate Change (Intergovernmental Panel on

Climate Change 2014) and the assessment on ecosystem services prepared by the Millennium Ecosystem Assessment (Millennium Ecosystem Assessment 2005). In recent years, however, scientific assessments have become a prominent feature of the landscape of policy-making in many areas. In some cases, assessments include projections regarding future states of important Earth systems (e.g., the climate system). But it is essential to bear in mind that such projections do not amount to predictions in any meaningful sense of the term (Bolin 1997). The error bars surrounding the IPCC's projections regarding climate change, for example, have gotten larger in some areas over time. Even then, actual developments regularly fall outside the envelope defined by the error bars surrounding IPCC projections.

Simulations, by contrast, are thought experiments based on models of systems that can be launched from some set of initial conditions and run repeatedly to explore the dynamics of the systems and, more specifically, to analyze their sensitivity to initial conditions. In many cases (e.g., the general circulation models of the Earth's climate system), the models are limited to biophysical conditions; they are useful in assessing the resilience of the system and the nature of thresholds and trigger mechanisms that could produce nonlinear and directional changes that would present challenges to policy-makers. More recently, modelers have endeavored to add a human component to such exercises, using what are known as agent-based models to explore the dynamics of socioecological systems (Gilbert 2008). Like assessments, simulations do not generate predictions that policy-makers can use in evaluating the consequences of the options available to them. But simulations are particularly valuable in efforts to understand the dynamics of complex systems in which telecouplings, nonlinear changes, and emergent properties give rise to developments that are apt to take policy-makers by surprise. In effect, their role is to raise awareness of the dynamics of complex systems rather than to tell policy-makers what choices to make in the face of fundamental uncertainties.

Going a step further, scenarios provide a means of thinking systematically about plausible futures in situations where uncertainty is high and there is little scope for forecasting. As a recent review essay puts it, scenarios "… are essentially descriptions of possible future events, typically constrained only be the condition of plausibility, and considered in terms of their implications for present-day decisions and actions" (Leschine et al.

2015: 646). Widely used in business settings, scenarios provide a means of helping people to break out of a business-as-usual mentality and to engage in innovative thinking about situations that involve more or less radical departures from the status quo but that are at the same time plausible. The value of scenario exercises lies not in their ability to anticipate what will happen but rather in their capacity to stimulate thinking about alternative futures. The underlying assumption is that those who are comfortable in moving beyond entrenched habits of thinking will be in a better position to respond agilely or nimbly to changing circumstances that call for the restructuring of prevailing governance systems.

Useful as these tools may be in helping policy-makers to identify options and to think about their probable consequences systematically, however, it is essential to bear in mind that the policy-makers themselves must take ultimate responsibility for making critical choices. In the nature of things, governance systems created to address societal problems frequently fail to live up to expectations, especially when we consider the pitfalls associated with implementation or efforts to move institutional arrangements from paper to practice. Given the nature of the complex systems we are dealing with in the Anthropocene, it is reasonable to expect that the incidence of governance failures will be particularly high in this era. Under the circumstances, it makes sense to adopt an experimental approach to the development of governance systems, making use of pilot projects where feasible, and building in mechanisms for assessing the results closely and for making mid-course corrections when it becomes clear that arrangements introduced to address particular needs for governance are not going to produce the intended results.

Taking Advantage of New Technologies

While governing complex systems requires a capacity to grasp the dynamics of large interconnected phenomena and to respond nimbly or agilely to nonlinear and surprising changes, new technologies are becoming available that can augment our capacity to deal with such challenges. New technologies are almost always double-edged swords. Advanced information systems, for example, allow us to understand the dynamics of complex systems in real time, but they also make possible forms of surveillance and invasions of privacy that are deeply troubling in ethical terms. Nevertheless, technological advances will play key roles in allowing policy-makers to meet the challenges of governing complex systems under conditions prevailing in the Anthropocene. Let me illustrate this potential through a commentary on the role of satellite observations in solving environmental problems.

As the examples compiled in a recent report from the Committee on Earth Observation Satellites make clear, satellite observations can play a number of roles that are relevant to governing complex systems (Committee on Earth Observation Satellites 2015). They can identify large-scale emerging problems that are difficult to detect through conventional means, as in the cases of the seasonal thinning of the stratospheric ozone layer and the massive destruction of tropical forests in remote areas like the Amazon Basin or Borneo. They can monitor changes in large biophysical systems in real time, as in the cases of the recession and thinning of sea ice in the Arctic, increased melting on the surface of the Greenland ice sheet, and the expansion of desert conditions in northwestern China. They can also assist efforts to deal with violations of the terms of environmental governance systems, as in the detection of ships engaged in illegal, unreported, and unregulated (IUU) fishing, or logging companies engaged in illegal timber harvesting in remote and sparsely populated areas.

Satellite observation systems have several attributes that make them particularly helpful in efforts to govern large complex systems. They can make observations covering extensive areas and feed them into databases that can be interrogated electronically in search of patterns of interest to those concerned with teleconnections, nonlinear processes, and the emergent properties of complex systems. They can make frequent observations (often several times a day or even every few hours) of large systems that may be moving toward thresholds likely to trigger sudden cascades of change that are dangerous to human well-being. Such systems have the capacity, for example, to provide early warning regarding tidal waves caused by earthquakes at sea or typhoons and to allow for more accurate monitoring of the tracks of hurricanes likely to make landfall in densely populated areas. On a more positive note, Earth observation systems can monitor the success of efforts to address large-scale environmental problems, including initiatives designed to reduce seasonal thinning of the stratospheric ozone layer or to reduce the intensity of air pollution that has been shown to have severe and widespread impacts on human health.

The fact that such contributions are technically feasible, however, does not mean that we will or even should make extensive use of them in efforts to govern complex systems in the Anthropocene. There may be paralyzing disagreements over who should pay for the launching and operation of the relevant satellites. Interagency turf battles may stymie the development of space programs, even when there is agreement that such activities are appropriate for inclusion in public sector activities. Powerful actors (e.g., companies relying on coal-fired power plants to produce electricity) may find that it is in their interests to block efforts to develop effective satellite observation systems to identify and track the behavior of major pollutants. Authoritarian governments may find that widespread dissemination of information about pollutants (e.g., PM 2.5 in China) makes it increasingly difficult to maintain control over their populations. What is more, there are significant ethical issues relating both to the acquisition of satellite observations and to the dissemination of these data to the public. Some may see a danger that satellite observations will become a form of surveillance, infringing the rights of various actors to privacy or that they will be used for dual purposes, collecting sensitive intelligence data at the same time as they are gathering information useful in addressing large-scale environmental problems. Others may object to policies of free and open access to the data collected, either on the grounds that some of the data are not suitable for general public consumption or on the grounds that such policies may undermine opportunities for private businesses to engage in profitable ventures centered on the acquisition and dissemination of Earth observations.

What this suggests is that there is a need to supplement or restructure the international regime dealing with satellite observations. The existing Group on Earth Observations (GEO) is a voluntary partnership consisting mainly of providers of Earth observation data and dedicated to improving the technical capacity of programs engaged in making satellite observations (Group on Earth Observations 2015). It appears to have performed well in these terms. An enhanced regime could deal with a number of important issues, including the licensing of operators, conditions designed to prevent the use of Earth observation systems for unauthorized purposes, rules governing access to data included in the resultant databases, the treatment of space debris, and the maintenance of this regime on an ongoing basis. One interesting question concerns the extent to which there is a need for some supranational body authorized to address issues relating to the harmonization of Earth observation data collected by a number of national and regional agencies (e.g., the European Space Agency, the National Aeronautics and Space Administration in the United States, and the Japan Aerospace Exploration Agency). The development of an effective international regime to address these issues would require a concerted effort to come to grips with a number of complex and potentially contentious issues. But the opportunities are substantial, and there is already a well-developed global community of people who are active in this realm. There is much to be said for moving forward with the development of such a regime from the perspective of maximizing the effectiveness of efforts to govern complex systems in the Anthropocene.

The Science/Policy Interface

Given what I have said about research opportunities and policy implications, what is the proper relationship between the research community and the policy community in dealing with the challenge of governing complex systems in the Anthropocene? One legitimate response to this question is that scientists and members of the attentive public more generally should avoid being drawn into the realm of policy-making regarding efforts to govern complex systems. Drawing a distinction between what he calls "horizons of necessity" and "horizons of feasibility," Richard Falk argues that we need a fundamental transformation of human consciousness to come to terms with the challenge of living well in the Anthropocene, and that this will require the development of an effective social movement that can break the power of business as usual. As he puts it, the "gap between feasibility and necessity cannot be closed ... except by a revolutionary or nonincremental jump that is transformational as far as feasibility is concerned, and ... this will not occur without a post-Marxist social mobilization from below" (Falk 2016: 101). This will require "engaged citizenship" dedicated to the development of a new discourse rather than the application of expert knowledge "reinforced by careful analysis and reasoned policy assessments" (Falk 2016: 98–99). From this perspective, scientists and thought leaders who engage with members of the policy community are more likely to fall victim to co-optation than to make a significant difference in breaking the hold of business as usual.

I sympathize with this perspective. It may well be that solving problems like climate change, the loss of biological diversity, and transboundary air pollution will require fundamental changes in human values and in the terms of the social contract governing relations between humans and the planet's biophysical systems. Nevertheless, the initiation, much less the completion, of this transformation may take some time. Falk himself reaches "the unhappy conclusion that the will of the species to survive is rather weak, that is, the dangers of catastrophic happenings induce what might be described as a new normal rather than a social demand to minimize risks" (Falk 2016: 101–102). With all due respect to the real danger of cooptation, therefore, I believe there is a strong case for encouraging engagement between the science community and the policy community in a collaborative effort to build the intellectual and social capital needed to meet the challenge of governing complex systems.

The first thing to note in this context is that starting with a sharp dichotomy between analysis and praxis regarding such matters is simplistic. It is true that scientists are motivated largely by the search for new knowledge and by the rewards of the academic world, while policy-makers are motivated by the pursuit of the interests of their governments and the prospect of public acclaim. Yet this is not the whole story. Those of us whose careers lie largely in the scientific community have a strong interest in efforts to apply our knowledge to address real-world concerns like climate change or the loss of biological diversity. Those whose careers lie largely in the policy community not only benefit from the growth of scientific knowledge, they also frequently contribute to the growth of knowledge regarding the creation of governance systems and the determinants of the effectiveness of these systems.

What does make sense is to think of the stages of the policy cycle in assessing the comparative advantages of scientists and policy-makers in addressing the governance of complex systems. Scientists have a large role to play in the generative phase of the cycle, identifying emerging problems and helping to shape the intellectual capital we can draw on in framing and addressing these problems. Partly, this is a matter of identifying new problems and improving understanding of the policy implications of these problems. The work of the scientists who have produced a succession of assessment reports under the auspices of the Intergovernmental Panel on Climate Change provides a dramatic illustration of this point. In part, however, it is a matter of enriching the toolkit of governance options available to address new problems. What I described earlier in this chapter as transcending the regulatory paradigm is an important case in point. With all due respect to the significance of regulatory concerns, it is important to realize that there are other ways to think about governance systems and that these alternatives may prove more effective in dealing with some problems than the standard operating procedures of regulatory politics. What I have characterized as principled governance and goal-setting in contrast to rule-making are good illustrations. But there are other alternatives to the regulatory paradigm that deserve consideration as well. Policy-makers, seeking to hammer out the specific language of mutually acceptable agreements in diplomatic settings, seldom have the luxury of being able to contemplate such matters in a concentrated manner. Yet, as we face the novel challenges of governing complex systems in the Anthropocene, they have a compelling need for access to new ways of thinking that offer the prospect of breaking the bonds of business as usual.

When it comes to the decision or choice stage of the policy cycle, on the other hand, members of the policy community have a clear comparative advantage. Negotiating the specific provisions of international agreements requires an immersion in the process and an attention to detail that is seldom possible for members of the science community. What is at stake here is the difference between cognitive leadership and entrepreneurial leadership (Young 1991). Whereas the cognitive leader influences the process by generating and disseminating new ways of thinking about the relevant problem, the entrepreneurial leader is an expert in seeing and seizing opportunities to make deals that will allow multiple actors to join forces in supporting the progressive development of governance systems and crafting textual provisions accordingly. The creation of successful governance systems always requires the combination of these distinct contributions. And the challenge of finding ways to address the requirements of governing complex systems simply reinforces the need for collaboration along the science/policy interface.

Moving governance systems from paper to practice and administering them on a day-to-day basis over time, by contrast, calls for talents that are in short supply in both the science community and the policy community. What is needed are the talents of professional administrators who understand the importance of adhering to well-defined procedures on the one hand but who are nimble or agile enough to avoid the onset of institutional arthritis on the other. An ability to transcend the rigidities of bureaucratic processes is especially important in governing complex systems where an ability to adjust nimbly to nonlinear and often surprising changes is essential. It will not suffice to endow governance systems with a capacity to adapt easily to changing circumstances if the bureaucratic propensities of those responsible for administering these systems make them excessively sluggish in practice. In this connection, it may help for members of the science community and the policy community to join forces in highlighting the novel features of governing complex systems, and emphasizing the importance of flexible administration as an essential feature of solving the problem of fit in such settings.

When it comes to assessing the effectiveness of governance systems and arriving at conclusions about how to make adjustments to improve their performance going forward, the need for collaboration along the science/policy interface is clear. The critical issue here arises from causal complexity and the importance of understanding the interactions among multiple drivers in pursuing sustainability in the Anthropocene. What is the role of defective institutional arrangements versus changes in biophysical forces (e.g., climate change) in explaining the fate of the cod stocks of the Northwest Atlantic? What is the role of HFCs in the climate equation, and can we make use of the Montreal Protocol to phase out the use of HFCs, even though they are not ozone-depleting substances? Questions of this sort come into focus repeatedly as we seek to address the governance of complex systems in which nonlinear change is a common event and emergent properties produce surprises regularly. The good news is that seeking to answer such questions requires input from both scientists and policy-makers and provides an obvious opportunity for building cooperative relationships between members of the two communities.

The Last Word

I began this book by emphasizing the problem of fit and arguing that complex systems have a number of attributes that will require the introduction of reconfigured institutional arrangements to achieve sustainability.

This does not mean that we need to set aside our existing intellectual capital regarding the governance of human-environment relations or abandon the social capital we have accumulated in this realm. But living in the Anthropocene does demand an ability to transcend business as usual and break the bonds of path dependence. I have tried to initiate this process, exploring the role of goal-setting as a governance strategy, introducing the idea of principled governance, and looking into the contributions of good governance. Yet these ideas are only a first installment on what can and should become a growth area among analysts and practitioners alike. Whether the pace of development of new social capital will prove adequate to meet the rising challenges of the Anthropocene is an open question. But there is no doubt about the importance of thinking hard about the conundrum of governing complex systems.

Notes

Introduction

1. The idea of the adaptive cycle directs attention to oscillations in which growth leads to breakdown or collapse followed by reorganization leading to renewed growth. So long as a system continues to follow this cyclical pattern, it is regarded as resilient (Gunderson and Holling 2002).

2. For a forceful recent reminder that markets cannot function at all, much less produce socially desirable outcomes, in the absence of well-developed rules of the game, see Reich 2015.

3. Recent contributions to the scientific literature suggest that this pause may have been no more than an artifact of statistical procedures used in analyzing trends in temperatures (Karl et al. 2015); 2015 was the hottest year on record, at least since the beginning of reasonably dependable observations.

4. True uncertainty, in contrast to risk, occurs in situations where there is no objective way to calculate the likelihood of a range of possible outcomes. In extreme cases, it may not be feasible even to identify the full range of possible outcomes.

Chapter 1

1. For information on IASC, consult the association's website: http://www.iasc-commons.org.

2. Consensus within the research community regarding such matters is also a worthy goal. But my main point here concerns the importance of precision and consistency.

3. Of course, there is typically a gap between the ideal and the actual with regard to the performance of governance systems. Like market failures, governance failures are common occurrences, even when governance systems seem well-crafted on paper.

4. Some analysts use the term resource regime to refer to situations involving natural resources, reserving the term environmental regime to apply to situations featuring environmental protection. But there is no consensus regarding this

distinction. I use the phrase environmental and resource regimes to cover all arrangements relating to the governance of human-environment relations.

5. In cases where the parties do not share a common understanding of the problem(s) to be solved or regime formation is a political gesture that has little to do with problem-solving, this conception of effectiveness will not apply. Properly speaking, therefore, my focus on effectiveness applies to a subset of the overall category of environmental and resource regimes.

6. There are cases in which an actor may choose to opt out of the social group to which an institution applies. This is one reason why some individuals choose to emigrate from their countries of origin and some states refuse to sign or ratify international agreements.

7. An important theme in regime analysis concerns the roles that collective entities (e.g., states) play as leaders and laggards in the formation and implementation of regimes. This has not been a focus of my work in this field.

8. This distinction bears a resemblance to what March and Olsen describe as the logic of consequences and the logic of appropriateness (March and Olsen 1998; Young 2001a).

9. The ozone regime has produced greater reductions in greenhouse gas emissions than the climate regime itself (Velders et al. 2007).

Chapter 2

1. A more radical view asserts that the existing approach to environmental governance, which acknowledges the state as the dominant actor, emphasizes distinct initiatives at different levels of social organization, and focuses on intergovernmental agreements at the international level, is fundamentally flawed and bound to fail (Park, Conca, and Finger 2008).

2. There are dissenters even in these cases. Some see evidence that countries have reduced emissions of chlorofluorocarbons voluntarily and point to the fact that healing the ozone layer will take decades (Murdoch and Sandler 1997). Others emphasize the role of non-regime factors in the effort to clean up the Rhine and comment on the slow pace of negotiations in this case (Bernauer 1996; Bernauer and Moser 1996).

3. The 2015 Paris Agreement on climate may make a difference over time. But there is little basis at this stage for treating the climate regime as a success.

Chapter 3

1. For an account that is skeptical about the comparability of biophysical systems and socioeconomic systems in this regard, see Nuttall 2012.

2. Some analysts draw a distinction between bifurcations, which are truly irreversible, and oscillating systems, which shift back and forth between two or more distinct states (Lenton 2012). In this analysis of social tipping points, I focus

on bifurcations (Scheffer 2009). Nevertheless, many real-world cases do not fall neatly into one or the other of these categories. A system may move back toward a preexisting state with regard to some of its attributes but not others.

3. To be precise, the armistice called for the fighting to stop at the eleventh hour of the eleventh day of the eleventh month of 1918.

4. For an analysis that takes a very long-term perspective on capitalist economies, see Piketty 2014.

5. To this day, New Orleans' defenses remain inadequate to deal with a hurricane as intense as Katrina.

6. The quoted phrase is from the UN Framework Convention on Climate Change, Art. 2.

7. There is a lack of consensus regarding the date of the onset of the Little Ice Age. Some analysts push the start of this event back to the beginning of the fourteenth century.

Chapter 4

1. For example, animals often sense oncoming storms and take steps to protect themselves from their impacts.

2. In many cases, the actual costs of solving environmental problems have proven to be substantially less than the projected or anticipated costs. But there is no guarantee that this will occur in the case of climate change.

3. For information on various applications of Fishbanks, see https://mitsloan.mit .edu/LearningEdge/simulations/fishbanks/Pages/fish-banks.aspx.

Chapter 5

1. A prominent case in point is the 1946 International Convention for the Regulation of Whaling.

2. On the effectiveness of regulatory arrangements, see Young 2011 and chapter 2 *supra*.

3. Recall, in this connection, Schelling's example of an actor who voluntarily agrees to pay a fine in the event of failure to comply as a means of making its commitment credible to others (Schelling 1960).

4. In America, environmental NGOs regularly send "bills" for membership renewals, even when they are really soliciting voluntary contributions.

5. Under the circumstances, it is no accident that the phrase "transforming our world" is now in use to characterize the sustainable development agenda for 2016–2030.

6. For enthusiastic accounts of progress in this realm, see United Nations 2015 (the final UN report on progress toward fulfillment of the MDGs) and Sachs 2015: chapter 14. It is important to note, however, that neither of these accounts

takes on the challenging task of demonstrating causal connections in discussing the performance of the MDGs.

7. For an extended discussion of the idea of good governance see chapter 7 *infra*.

8. SDG Goal 13 deals with climate change. But a footnote makes it clear that the UNFCCC provides the appropriate mechanism for addressing this problem.

Chapter 6

1. The Earth Charter is a product of civil society finalized in 2000 rather than a UN document (Earth Charter 2002).

2. Principle 2 of the Rio Declaration augments Principle 21 of the Stockholm Declaration by adding the word developmental to the phrase "environmental and developmental policies."

3. China is now the leading emitter of carbon dioxide, accounting for ~27 percent of the global total. Not counting the EU as a single entity, India is third, accounting for ~7 percent.

4. Readers will recognize the influence of this idea on the Brundtland Commission's definition of sustainable development (World Commission on Environment and Development 1987).

5. For a seminal statement of this view, see Singer 1997; a critical appraisal appears in Herscovici 1985.

6. On the idea of rights as trumps that can outweigh conventional calculations of benefits and costs, see Dworkin 1978.

Chapter 7

1. For an analysis of sustainable development that also stresses the importance of these factors, see Sachs 2015, chapter 14.

2. For further information, consult the association's website, http://c.ymcdn.com/ sites/www.iap2.org/resource/resmgr/Foundations_Course/IAP2_P2_Spectrum .pdf.

3. On the decision rules in operation in COP/MOP settings, see Breitmeier, Young, and Zürn 2006, chapter 4.

4. For more information on TI, consult their website, http://www.transparency .org/cpi.

5. The US now argues that many of the provisions of UNCLOS are legally in force because they have acquired the status of customary international law, despite the refusal of the Senate to ratify the convention.

6. For an analysis that extends this discussion to what it calls private transnational regulatory organizations (PTROs) as well as the intergovernmental organizations established to administer international regimes, see Abbott, Green, and Keohane (2016).

7. Even in the US, some executive agreements are treated as legally significant instruments.

8. See chapter 2 *infra* for a discussion of the distinction among outputs, outcomes, and impacts.

9. Soon after taking office in 2001, for example, the Bush Administration renounced the 1997 Kyoto Protocol signed (but not ratified) by the US under the Clinton Administration.

10. For a spectacular example, see James Risen's account of the disappearance of billions of US dollars shipped to Iraq for the ostensible purpose of restarting the Iraqi economy following the fall of Saddam Hussein (Risen 2014).

11. For an argument that environmental conditions may deteriorate in specific cases even when accountability is high or rising, see Kramarz and Park (2016).

12. I should note, however, that Ronald Mitchell has engaged in a sustained effort to address this problem. See http://iea.uoregon.edu.

Conclusion

1. The Global Carbon Project has reported the good news that emissions of carbon dioxide appear to have leveled off or even dipped slightly in 2015 (Global Carbon Project 2015). But note that even if this turns out to be the beginning of a trend, concentrations of CO_2 in the Earth's atmosphere will continue to rise.

References

Abbott, K. W., J. E. Green, and R. O. Keohane. 2016. "Organizational Ecology and Institutional Change in Global Governance." *International Organization* 70:247–277.

Akerlof, G. A., and R. J. Shiller. 2009. *Animal Spirits: How Human Psychology Drives the Economy, and Why It Matters for Global Capitalism.* Princeton: Princeton University Press.

Alley, R. 2000. *The Two-Mile Time Machine: Ice Cores, Abrupt Climate Change, and Our Future.* Princeton: Princeton University Press.

AMAP. 1997. *State of the Arctic Environment Report.* Oslo: AMAP.

Andersen, S. O., and K. M. Sarma. 2012. *Protecting the Ozone Layer: The United Nations History.* New York: Earthscan.

Andresen, S., and J. Wettestad. 2004. "Case Studies of the Effectiveness of International Environmental Regimes," 27–48. In A. Underdal and O. R. Young, eds., *Regime Consequences.* Dordrecht: Kluwer Academic Publishers.

Andresen, S., E. L. Boassum, and G. Hønneland, eds. 2012. *International Environmental Agreements: An Introduction.* London: Routledge.

Archer, D., and S. Rahmstorf. 2010. *The Climate Crisis: An Introductory Guide to Climate Change.* Cambridge: Cambridge University Press.

Arctic Climate Impact Assessment (ACIA). 2004. *Impacts of a Warming Arctic.* Cambridge: Cambridge University Press.

Arctic Human Development Report (AHDR). 2004. *Arctic Human Development Report.* Akureyri, Iceland: Stefansson Arctic Institute.

Ariely, D. 2010. *Predictable Irrationality: The Hidden Forces that Shape Our Decisions.* New York: Harper Perennial.

Aspen Institute. 2011. *The Shared Future: A Report of the Aspen Institute Commission on Arctic Climate Change.* Washington, DC: Aspen Institute.

Axelrod, R. 1984. *The Evolution of Cooperation.* New York: Basic Books.

Ba, A. D., and M. J. Hoffmann, eds. 2005. *Contending Perspectives on Global Governance: Coherence, Contestation, and World Order.* London: Routledge.

Barrett, S. 2003. *Environment and Statecraft: The Strategy of Environmental Treaty Making.* New York: Oxford University Press.

Barrett, S. 2007. *Why Cooperate? The Incentive to Supply Collective Goods.* Oxford: Oxford University Press.

Benedick, R. E. 1998. *Ozone Diplomacy: New Directions in Safeguarding the Planet.* Cambridge, MA: Harvard University Press.

Benedict, R. 2006. *The Chrysanthemum and the Sword.* New York: Mariner Books.

Berkes, F., J. Codding, and C. Folke, eds. 2003. *Navigating Social-Ecological Systems—Building Resilence for Complexity and Change.* Cambridge: Cambridge University Press.

Bernauer, T. 1996. "Protecting the Rhine River against Chloride Pollution," 201–232. In R. O. Keohane and M. Levy, eds., *Institutions for Environmental Aid.* Cambridge, MA: MIT Press.

Bernauer, T., and P. Moser. 1996. "Reducing Pollution of the Rhine River: The Influence of International Cooperation." *Journal of Environment & Development* 5:391–417.

Bernauer, T., and T. Siegfried. 2008. "Compliance and Enforcement in International Water Agreements: The Case of the Naryn/Syr Darya Basin." *Global Governance* 14:479–501.

Bernstein, S. 2005. "Legitimacy in Global Environmental Governance." *Journal of International Law and International Relations* 1:139–166.

Bernstein, S. 2013. "Rio+20: Sustainable Development in a Time of Multilateral Decline." *Global Environmental Politics* 13:12–21.

Bernstein, S., and L. W. Pauly, eds. 2007. *Global Liberalism and Political Order: Towards a New Grand Compromise.* Albany: State University of New York Press.

Betsill, M., and E. Corell, eds. 2007. *NGO Diplomacy: The Influence of Nongovernmental Organizations in International Environmental Negotiations.* Cambridge, MA: MIT Press.

Biermann, F. 2007. "'Earth System Governance' as a Crosscutting Theme of Global Change Research." *Global Environmental Change* 17:326–337.

Biermann, F. 2014. *Earth System Governance: World Politics in the Anthropocene.* Cambridge, MA: MIT Press.

Biermann, F., and S. Bauer, eds. 2005. *A World Environment Organization: Solution or Threat for Effective International Environmental Governance.* Aldershot, UK: Ashgate.

Biermann, F., et al. 2009. *Earth System Governance: People, Places and the Planet.* IHDP Report No. 20. Bonn: IHDP.

Biermann, F., and B. Siebenhüner, eds. 2009. *Managers of Global Change: The Influence of International Environmental Bureaucracies.* Cambridge, MA: MIT Press.

Biermann, F., and P. Pattberg, eds. 2012. *Global Environmental Governance Reconsidered.* Cambridge, MA: MIT Press.

Bloom, D. E. 2011. "7 Billion and Counting." *Science* 333:562–569.

Bolin, B. 1997. "Scientific Assessment of Climate Change," 83–109. In G. Fermann, ed., *International Politics of Climate Change: Key Issues and Critical Actors*. Oslo: Scandinavian University Press.

Boyd, E., and C. Folke. 2011. *Adapting Institutions: Governance, Complexity, and Social-Ecological Systems*. Cambridge: Cambridge University Press.

Bratspies, R. M., and R. A. Miller, eds. 2006. *Transboundary Harm in International Law: Lessons from the Trail Smelter Arbitration*. Cambridge: Cambridge University Press.

Breitmeier, H., O. R. Young, and M. Zürn. 2006. *Analyzing International Environmental Regimes: From Case Study to Database*. Cambridge, MA: MIT Press.

Breitmeier, H., A. Underdal, and O. R. Young. 2011. "The Effectiveness of International Environmental Regimes: Comparing and Contrasting Findings from Quantitative Research." *International Studies Review* 13:1–27.

Brennan, D. G., ed. 1961. *Arms Control, Disarmament, and National Security*. New York: George Braziler.

Brennan, G., and J. M. Buchanan. 1984. *The Reason of Rules: Constitutional Political Economy*. Cambridge: Cambridge University Press.

Bromley, D. 2012. "Environmental Governance as Stochastic Belief Updating: Crafting Rules to Live By." *Ecology and Society* 17 (3), essay 14.

Brondisio, E., E. Ostrom, and O. R. Young. 2009. "Social Capital and Ecosystem Services: Institutions and Governance at Multiple Levels." *Annual Review of Environment and Resources* 34:253–278.

Brousseau, E., et al., eds. 2012. *Global Environmental Commons*. Oxford: Oxford University Press.

Brown Weiss, E. 1993. "International Environmental Issues and the Emergence of a New World Order." *Georgetown Law Journal* 81:675–710.

Brown Weiss, E. 1995. "Environmental Equity: The Imperative for the Twenty-First Century," 17–27. In W. Land, ed., *Sustainable Development and International Law*. London: Graham and Trotman/Martinus Nijhoff.

Brown Weiss, E., and H. K. Jacobson, eds. 1998. *Engaging Countries: Strengthening Compliance with International Environmental Accords*. Cambridge, MA: MIT Press.

Brown, L. 1978. *The Twenty Ninth Day*. New York: W.W. Norton.

Brown, L. R., G. Gardner, and B. Halweil. 1998. *Beyond Malthus: Sixteen Dimensions of the Population Problem, Worldwatch Paper No. 143*. Washington, DC: Worldwatch Institute.

Brundtland, G. H. 2005. *Madam Prime Minister: A Life in Power and Politics*. New York: Farrar, Straus and Giroux.

Bryner, G. 1995. *Blue Skies, Green Politics: The Clean Air Act of 1990 and Its Implementation*. Washington, DC: Congressional Quarterly Books.

Bull, H. 1977. *The Anarchical Society*. New York: Columbia University Press.

Buzan, B. 2004. *From International to World Society*. Cambridge: Cambridge University Press.

Carpenter, S. 2003. *Regime Shifts in Lake Ecosystems: Pattern and Variation. Excellence in Ecology Series 15. Oldendorf/Luje*. Oldendorf/Luhe, Germany: Ecological Institute.

Carpenter, S. R., and K. L. Cottingham. 2002. "Resilience and the Restoration of Lakes," 51–70. In L. H. Gunderson and L. Pritchard, eds., *Resilience and the Behavior of Large-Scale Systems*. Washington, DC: Island Press.

Carpenter, S., et al. 2009. "Science for Managing Ecosystem Services: Beyond the Millennium Ecosystem Assessment." *Proceedings of the National Academy of Sciences of the United States of America* 106:1305–1313.

Cash, D. W., et al. 2006. "Scale and Cross-Scale Dynamics: Governance and Information in a Multilevel World." *Ecology and Society* 11:181–192.

Chapin, F. S., G. P. Kofinas, and C. Folke, eds. 2009. *Principles of Ecosystem Stewardship: Resilience-Based Natural Resource Management in a Changing World*. New York: Springer.

Chapin, F. S., M. Sommerkorn, M. D. Robards, and K. Hillmer-Pegram. 2015. "Ecosystem Stewardship: A Resilience Framework for Arctic Conservation." *Global Environmental Change* 34:207–217.

Chasek, P., and L. H. Wagner, eds. 2012. *The Roads from Rio: Lessons Learned from Twenty Years of Multilateral Environmental Negotiations*. New York: RFF Press.

Chayes, A., and A. H. Chayes. 1995. *The New Sovereignty: Compliance with International Regulatory Agreements*. Cambridge, MA: Harvard University Press.

Clark, W. C. 2007. "Sustainability Science: A Room of Its Own." *Proceedings of the National Academy of Sciences of the United States of America* 104:1737–1738.

Coase, R. 1960. "The Problem of Social Cost." *Journal of Law & Economics* 3:1–44.

Coicaud, J. M., and D. Warner, eds. 2013. *Ethics and International Affairs: Extent and Limits*. 2nd ed. Tokyo: UNU Press.

Cole, D. H. 2002. *Pollution and Property: Comparing Ownership Institutions for Environmental Protection*. Cambridge: Cambridge University Press.

Collier, P. 2008. *The Bottom Billion: Why the Poorest Countries are Failing and What Can Be Done about it*. Oxford: Oxford University Press.

Comacho, D. E., ed. 1998. *Environmental Injustices, Political Struggles: Race, Class, and the Environment*. Durham, NC: Duke University Press.

Committee on Earth Observation Satellites (CEOS). 2015. *Applications of Satellite Earth Observations: Serving Society, Science, and Industry*. Tokyo: JAXA.

Commoner, B. 1971. *The Closing Circle: Nature, Man and Technology*. New York: Knopf.

Conca, K. 2006. *Governing Water: Contentious Transnational Politics and Global Institution Building*. Cambridge: MIT Press.

Connolly, B. 1996. "Increments for the Earth: The Politics of Environmental Aid," 327–365. In R. O. Keohane and M. A. Levy, eds., *Institutions for Environmental Aid*. Cambridge, MA: MIT Press.

Copenhagen Accord. 2009. "Outcome Document from UNFCCC COP15." Text available at: http://unfccc.int.

Cornell, S. E., I. C. Prentice, J. I. House, and C. J. Downy, eds. 2012. *Understanding the Earth System: Global Change Science for Application*. Cambridge: Cambridge University Press.

Costello, C., S. D. Gaines, and J. Lynham. 2008. "Can Catch Shares Prevent Fisheries Collapse?" *Science* 321:678–681.

Crowder, L., et al. 2006. "Resolving Mismatches in U.S. Ocean Governance." *Science* 313:617–618.

Crutzen, P. 2002. "Geology of Mankind—The Anthropocene." *Nature* 415:23.

Crutzen, P., and E. F. Stoermer. 2002. "The Anthropocene." *IGBP Newsletter* 41:17–18.

Dahl, Robert A. 1961. *Who Governs?* New Haven: Yale University Press.

Dai, X. 2005. "Why Comply? The Domestic Constituency Mechanism." *International Organization* 59:363–398.

Dai, X. 2007. *International Institutions and National Policies*. Cambridge: Cambridge University Press.

Davies, J. 2016. *The Birth of the Anthropocene*. Oakland: University of California Press.

Delmas, M., and O. R. Young, eds. 2009. *Governance for the Environment: New Perspectives*. Cambridge: Cambridge University Press.

Deutsch, K. W. 1963. *The Nerves of Government: Models of Political Communication and Control*. New York: The Free Press.

Diamond, J. 2005. *Collapse: How Societies Choose to Fail or Succeed*. New York: Viking.

Dietz, T., E. Ostrom, and P. C. Stern. 2003. "The Struggle to Govern the Commons." *Science* 302:1907–1912.

Dowie, E. 1995. *Losing Ground: American Environmentalism at the Close of the Twentieth Century*. Cambridge, MA: MIT Press.

Downie, D., and T. Fenge. 2003. *Northern Lights against POPs: Toxic Threats in the Arctic*. Montreal: McGill-Queens University Press.

Dryzek, J. 1990. *Discursive Democracy: Politics, Policy, and Political Science*. Cambridge: Cambridge University Press.

Dryzek, J. 2014. "Institutions for the Anthropocene: Governance in a Changing Earth System." *British Journal of Political Science*. doi:10.1017/S00071234000453.

Dupont, C. 1993. "Switzerland, France, Germany, The Netherlands," 97–115. In G. O. Foure and J. Z. Rubin, eds., *Culture and Negotiation: The Resolution of Water Disputes*. Newberry Park, CA: Sage Publications.

Duit, A., and V. Galaz. 2008. "Governance and Complexity—Emerging Issues for Governance Theory." *Governance: An International Journal of Policy, Administration and Institutions* 21:311–335.

Dworkin, R. 1978. *Taking Rights Seriously*. Cambridge, MA: Harvard University Press.

Earth Charter. 2002. Text available at: http://www.earthcharterinaction.org.

Ebbin, S., A. H. Hoel, and A. Sydnes, eds. 2005. *A Sea Change: The Exclusive Economic Zones and Governance Institutions for Marine Living Resources*. Dordrecht, the Netherlands: Springer Verlag.

Economist, The. 2011, "Welcome to the Anthropocene," 399 (8735): 13.

Egan, T. 2006. *The Worst Hard Time*. New York: Houghton Mifflin.

Ehrlich, P., and J. Holdren. 1971. "Impact of Population Growth." *Science* 171:1212–1217.

Enderlin, H., S. Wälti, and M. Zürn, eds. 2010. *Handbook on Multi-Level Governance*. Cheltenham: Edward Elgar.

English, J. 2013. *Ice and Water: Politics, Peoples, and the Arctic Council*. Toronto: Allen Lane.

Fagin, D. 2013. *Tom's River: A Story of Science and Salvation*. New York: Bantam.

Falk, R. 1971. *This Endangered Planet: Prospects and Proposals for Human Survival*. New York: Random House.

Falk, R. 2016. "Scholarship as Engagement on a New Earth," 97–114. In S. Nicholson and S. Jinnah, eds., *New Earth Politics*. Cambridge, MA: MIT Press.

Faude, B., and T. Gehring. (n.d.). "Institutional Ecology and the Evolution of Broader Structures in Regime Complexes: The Trade and Environment Overlap." Available from: Thomas.gehring@sowi.uni-bamberg.de.

Finnemore, M., and K. Sikkink. 1998. "International Norm Dynamics and Political Change." *International Organization* 52:887–917.

Finus, M., and S. Tjøtta. 2003. "The Oslo Protocol on Sulfur Reduction: The Great Leap Forward?" *Journal of Public Economics* 87:2031–2048.

Folke, C., et al. 2004. "Regime Shifts, Resilience, and Biodiversity in Ecosystem Management." *American Review of Ecology, Evolution and Systematics* 35:557–581.

Folke, C. 2006. "Resilience: The Emergence of a Perspective for Social-Ecological Systems Analyses." *Global Environmental Change* 16:268–281.

Franck, T. M. 1990. *The Power of Legitimacy among Nations*. New York: Oxford University Press.

Franck, T. M. 1995. *Fairness in International Law and Institutions*. New York: Oxford University Press.

Freestone, D., and E. Hay, eds. 1996. *The Precautionary Principle and International Law: The Challenge of Implementation*. The Hague: Kluwer Law International.

Friedheim, R., ed. 2001. *Toward a Sustainable Whaling Regime*. Seattle: University of Washington Press.

Future Earth. n.d. Research Program, updates available at: http://www.futureearth.org.

Galaz, V., et al. 2008. "The Problem of Fit among Biophysical Systems, Environmental and Resource Regimes, and Broader Governance Systems: Insights and Emerging Challenges," 147–186. In O. R. Young, L. King, and H. Schroeder, eds. *Institutions and Environmental Change*. Cambridge, MA: MIT Press.

Galaz, V. 2014. *Global Environmental Governance, Technology and Politics: The Anthropocene Gap*. Cheltenham, UK: Edward Elgar.

Galaz, V., et al., eds. 2010. "Governance, Complexity, and Resilience," a special issue of *Global Environmental Change,* 20 (3).

Gibson, C. C., E. Ostrom, and T. K. Ahn. 2000. "The Concept of Scale and the Human Dimensions of Global Change." *Ecological Economics* 32:217–239.

Gilbert, N. 2008. *Agent-Based Models*. Los Angeles: Sage Publications.

Gladwell, M. 2002. *Tipping Point: How Little Things Can Make a Big Difference*. New York: Little Brown.

Gleick, P. H. 1998. *The World's Water: The Biennial Report on Freshwater Resources*. Washington, DC: Island Press.

Global Carbon Project. 2015. *Carbon Budget 2015*. www.globalcarbonproject.org.

Goodwin, D. K. 1994. *No Ordinary Time: Franklin and Eleanor Roosevelt: The Home Front in World War II*. New York: Simon and Schuster.

Goodwin, J., and J. M. Jasper, eds. 2009. *The Social Movement Reader*. Chichester, UK: Blackwell Publishing Ltd.

Greenhalgh, S. 2008. *Just One Child: Science and Policy in Deng's China*. Berkeley: University of California Press.

Griggs, D., et al. 2013. "Sustainable Development Goals for People and the Planet." *Nature* 495:305–307.

Group on Earth Observations (GEO). 2015. "A Summary Report of the 12th Plenary Session of the Group on Earth Observations (GEP-XII) and GEO 2015 Mexico City Summit," http://www.iisd.ca/geo/12.

Gruber, L. 2000. *Ruling the World: Power Politics and the Rise of Supranational Institutions*. Princeton: Princeton University Press.

Gunderson, L. H., and C. S. Holling, eds. 2002. *Panarchy: Understanding Transformations in Human and Natural Systems*. Washington, DC: Island Press.

Haas, P. M., R. O. Keohane, and M. L. Levy, eds. 1993. *Institutions for the Earth: Sources of Effective International Environmental Protection*. Cambridge, MA: MIT Press.

Harari, Y. N. 2015. *Sapiens: A Brief History of Humankind*. New York: Harper.

Hardin, G. 1968. "The Tragedy of the Commons." *Science* 162:1243–1248.

Harr, J. 1996. *A Civil Action*. New York: Vintage Books.

Hart, H. L. A. 1961. *The Concept of Law*. Oxford: Oxford University Press.

Hart, J. 1983. *The New International Economic Order: Conflict and Cooperation in North-South Economic Relations*. New York: St. Martin's Press.

Hayek, F. A. 1973. *Rules and Order*. Vol. 1. Law, Legislation, and Liberty. Chicago: University of Chicago Press.

Held, D. 1999. *Global Transformations*. Cambridge: Polity Press.

Helm, C., and D. Sprinz. 2000. "Measuring the Effectiveness of International Environmental Regimes." *Journal of Conflict Resolution* 44:639–652.

Herscovici, A. 1985. *Second Nature: The Animal Rights Controversy*. Montreal: CBC Enterprises.

High-Level Panel of Eminent Persons on the Post-2015 Development Agenda. 2013. *A New Global Partnership: Eradicate Poverty and Transform Economies Through Sustainable Development*. New York: United Nations Publications.

Hiltunen, H. 1994. *"Finland and Environmental Problems in Russia and Estonia," discussion paper*. Helsinki: Finnish Institute of International Affairs.

Holling, C. S., and H. L. Gunderson. 2002. "Resilience and Adaptive Cycles," 25–62. In L. H. Gunderson and C. S. Holling, eds., *Panarchy: Understanding Transformation in Human and Natural Systems*. Washington, DC: Island Press.

Homer-Dixon, T. 2006. *The Upside of Down: Catastrophe, Creativity and the Renewal of Civilization*. Washington, DC: Island Press.

Hovi, J., D. Spring, and A. Underdal. 2003a. "The Oslo-Potsdam Solution to Measuring Regime Effectiveness: Critique, Response, and the Road Ahead." *Global Environmental Politics* 3:74–96.

Hovi, J., D. Sprinz, and A. Underdal. 2003b. "Regime Effectiveness and the Oslo-Potsdam Solution: A Rejoinder to Oran Young." *Global Environmental Politics* 3:105–107.

Hurrell, A. 1992. "Brazil and the International Politics of Amazonian Deforestation," 398–429. In A. Hurrell and B. Kingsbury, eds., *The International Politics of the Environment*. Oxford: Clarendon Press.

Hurrell, A. 2007. *On Global Order: Power, Values, and the Constitution of International Society*. Oxford: Oxford University Press.

Intergovernmental Panel on Climate Change. 2007. *Fourth Assessment Report*. Text available at: http://www.ipcc.ch.

Intergovernmental Panel on Climate Change. 2014. *Fifth Assessment Report*. Text available at: http://www.ipcc.ch.

International Association for Public Participation. www.iap2.org.

Iudicello, S., M. Weber, and R. Wieland. 1999. *Fish, Markets, and Fishermen: The Economics of Overfishing*. Washington, DC: Island Press.

Jasanoff, S., ed. 2004. *States of Knowledge: The Co-production of Science and the Social Order*. New York: Routledge.

Janssen, M., ed. 2002. *Complexity and Ecosystem Management: The Theory and Practice of Multi-Agent Systems.* Cheltenham, UK: Edward Elgar.

Jinnah, S. 2010. "Overlap Management in the World Trade Organization: Secretariat Influence on Trade-Environment Politics." *Global Environmental Politics* 10:54–79.

Jinnah, S. 2014. *Post-Treaty Politics: Secretariat Influence in Global Environmental Governance.* Cambridge, MA: MIT Press.

Declaration, Johannesburg. 2002. "Outcome Document from the UN Conference on Sustainable Development." UN Doc. A/CONF.199/20.

Johnson, N. 2009. *Simply Complexity: A Clear Guide to Complexity Theory.* Oxford: Oneworld Publications.

Kahneman, D. 2011. *Thinking Fast and Slow.* New York: Farrar, Straus, and Giroux.

Kahneman, D., and A. Tversky. 2000. *Choices, Values, and Frames.* Cambridge: Cambridge University Press.

Kaldor, M. 2003. *Global Civil Society: An Answer to War.* Cambridge: Polity Press.

Kanie, N., et al. 2012. "A Charter Moment: Restructuring Governance for Sustainability." *Public Administration and Development* 32:292–304.

Kanie, N., and F. Biermann eds. 2017. *Governing through Goals: The Sustainable Development Goals and a New Governance Strategy in the 21st Century.* Cambridge: MIT Press.

Karabell, Z. 2014. *The Leading Indicators: A Short History of the Numbers that Rules the World.* New York: Simon and Schuster.

Karl, T. R., et al. 2015. "Possible Artifacts of Data Biases in the recent Global Surface Warming Hiatus." *Science* 348 (June 26): 1469–1472.

Keohane, R. O., and D. G. Victor. 2011. "The Regime Complex for Climate." *Perspectives on Politics* 9:7–23.

Keohane, R. O., and E. Ostrom, eds. 1995. *Local Commons and Global Interdependence.* London: Sage Publications.

Keynes, J. M. 1920. *The Economic Consequences of the Peace.* New York: Harcourt Brace Jovanovich.

Kingdon, J. W. 1995. *Agendas, Alternatives, and Public Policies.* 2nd ed. Boston: Addison-Wesley.

Kolata, G. 2001. *Flu: The Story of the Great Influenza Epidemic of 1918 and the Search for the Virus that Caused It.* New York: Touchstone.

Kolbert, E. 2014. *The Sixth Extinction: An Unnatural History.* New York: Henry Holt.

Kramarz, T., and S. Park. 2016. "Accountability in Global Environmental Governance: A Meaningful Tool for Action?" *Global Environmental Politics* 16:1–21.

Krasner, S. D., ed. 1983. *International Regimes.* Ithaca, NY: Cornell University Press.

Krasner, D. D. 1983. "Structural Causes and Regime Consequences: Regimes as Intervening Variables," 1–21. In S. D. Krasner, ed., *International Regimes*. Ithaca: Cornell University Press.

Laudato Si'. 2015. *Encyclical Letter Laudato Si' of the Holy Father Francis On Caring for Our Common Home* . www.papalencyclicals.net.

Leadership Council for the Sustainable Development Solutions Network. 2013. "An Action Agenda for Sustainable Development," Report to the UN Secretary General.

Lee, K. 1992. *Compass and Gyroscope*. Washington, DC: Island Press.

Leggett, J. 2001. *The Carbon War: Global Warming and the End of the Oil Era*. New York: Routledge.

Lenton, T. 2012. "Arctic Tipping Points." *Ambio* 41:10–22.

Lenton, T., et al. 2008. "Tipping Elements in the Earth's Climate System." *Proceedings of the National Academy of Sciences of the United States of America* 105:1786–1793.

Leopold, A. 1966. *A Sand County Almanac, with Essays on Conservation from Round River*. New York: Ballantine Books.

Leschine, T. M., et al. 2015. "What-If Scenario Modeling to Support Oil Spill Preparedness and Response Decision-Making." *Human and Ecological Risk Assessment: An International Journal* 21:646–666.

Levin, S. A. 1999. *Fragile Dominion—Complexity and the Commons*. Cambridge: Perseus Publishers.

Linden, E. 2006. *The Winds of Change: Climate, Weather, and the Destruction of Civilizations*. New York: Simon and Schuster.

Litfin, K. T. 1994. *Ozone Discourses: Science and Politics in Global Environmental Cooperation*. New York: Columbia University Press.

Liu, J., et al. 2013. "Framing Sustainability in a Telecoupled World." *Ecology and Society* 18 (2): 26. http://dx.doi.org/10.5751/.

Liu, J. et al. 2015. "Systems Integration for Global Sustainability," *Science* 347 (6225): 963–974. doi:10.1126/science.1258832.

Lovejoy, T., and L. Hannah, eds. 2005. *Climate Change and Biodiversity*. New Haven: Yale University Press.

Lovelock, J. E. 1979. *Gaia: A New Look at Life on Earth*. New York: Oxford University Press.

Lyon, T. F., and J. W. Maxwell. 2004. *Corporate Environmentalism and Public Policy*. Cambridge: Cambridge University Press.

Lyster, S. 1995. *International Wildlife Law*. Cambridge: Grotius Publishers.

Maier, P. 2011. *Ratification: The People Debate the Constitution, 1987–1788*. New York: Simon and Schusater.

March, J. G., and J. P. Olsen. 1998. "The Institutional Dynamics of International Political Orders." *International Organization* 52:943–969.

Matson, P., W. C. Clark, and K. Andersson. 2016. *Pursuing Sustainability: A Guide to the Science and Practice*. Princeton: Princeton University Press.

Mattoo, A., and A. Subramanian. 2012. "Equity in Climate Change: An Analytical Review." *World Development* 40:1083–1097.

Mayewski, P. A., and F. White. 2002. *The Ice Chronicles: The Quest to Understand Global Climate Change*. Hanover, NH: University Press of New England.

McKibben, B. 2013. *Oil and Honey: The Education of an Unlikely Activist*. Collingwood, Australia: Black Inc.

McNeill, J. R., and P. Engelke. 2014. *The Great Acceleration: An Environmental History of the Anthropocene since 1945*. Cambridge, MA: Harvard University Press.

Meadows, D. H. 2008. *Thinking in Systems*. White River Junction, VT: Chelsea Green.

Meadows, D. H., D. L. Meadows, J. Randers, and W. W. Behrens. 1972. *The Limits to Growth*. New York: Universe Books.

Mearsheimer, J. J. 1994/1995. "The False Promise of International Institutions." *International Security* 19:5–49.

Miles, E. L., A. Underdal, S. Andresen, J. Wettestad, J. B. Skjaerseth, and E. M. Carlin. 2002. *Environmental Regime Effectiveness: Confronting Theory with Evidence*. Cambridge, MA: MIT Press.

Millennium Ecosystem Assessment. 2005. *Ecosystems and Human Well-Being: Synthesis*. Washington, DC: Island Press.

Miller, M. A. L. 1995. *The Third World in Global Environmental Politics*. Boulder, CO: Lynne Rienner.

Mintzer, I. M., ed. 1992. *Confronting Climate Change: Implications and Responses*. Cambridge: Cambridge University Press.

Mirovitskaya, N. S., M. Clark, and R. G. Purver. 1993. "North Pacific Fur Seals: Regime Formation as a Means of Resolving Conflict," 22–55. In O. R. Young and G. Osherenko, eds., *Polar Politics: Creating International Environmental Regimes*. Ithaca, NY: Cornell University Press.

Mitchell, M. 2009. *Complexity: A Guided Tour*. Oxford: Oxford University Press.

Mitchell, R. B. 1994. *International Oil Pollution at Sea: Environmental Policy and Treaty Compliance*. Cambridge, MA: MIT Press.

Mitchell, R. B. 1996. "Compliance Theory: An Overview," 3–28. In J. Cameron, J. Worksman, and P. Roderick, eds., *Improving Compliance with International Law*. London: Earthscan Publications.

Mitchell, R. B. 2004. "A Quantitative Approach to Evaluating International Environmental Regimes," 121–149. In A. Underdal and O. R. Young, eds., *Regime Consequences*. Dordrecht: Kluwer Academic Publishers.

Mitchell, R. B. 2008. "Evaluating the Performance of International Environmental Institutions: What to Evaluate and How to Evaluate It?" 79–114. In O. R.

Young, L. King., and H. Schroeder, eds., *Institutions and Environmental Change*. Cambridge, MA: MIT Press.

Mitchell, R. B., et al. 2006. *Global Environmental Assessment: Information and Influence*. Cambridge, MA: MIT Press.

Müller, B., N. Höhne, and C. Ellermann. 2009. "Differentiating (Historic) Responsibilities for Climate Change." *Climate Policy* 9:593–611.

Murdoch, J. C., and T. Sandler. 1997. "The Voluntary Provision of a Public Good: The Case of Reduced CFC Emissions and the Montreal Protocol." *Journal of Public Economics* 63:331–349.

Murdoch, J. S., T. Sandler, and K. Sargent. 1997. "A Tale of Two Collectives: Sulphur versus Nitrogen Oxide Emissions Reduction in Europe." *Economica* 64:281–301.

National Research Council. 2015a. *Climate Intervention: Carbon Dioxide Removal and Reliable Sequestration*. Washington, DC: National Academies Press.

National Research Council. 2015b. *Climate Intervention: Reflecting Sunlight to Cool Earth*. Washington, DC: National Academies Press.

National Research Council. 2015c. "Arctic Matters: The Global Connection to Changes in the Arctic." http://nas-sites.org/arctic.

Nordhaus, W. 2008. *A Question of Balance: Weighing the Options in Global Warming Policies*. New Haven: Yale University Press.

Nuttall, M. 2012. "Tipping Points and the Human World: Living with Change and Thinking about the Future." *Ambio* 41:96–105.

Nye, J. S. 2011. *The Future of Power*. New York: Public Affairs.

Oberthür, S. 2009. "Interplay Management: Enhancing Environmental Policy Integration among International Institutions." *International Environmental Agreement: Politics, Law and Economics* 9:371–391.

Oberthür, S., and T. Gehring, eds. 2006. *Institutional Interaction in Global Environmental Governance: Synergy and Conflict among International and EU Policies*. Cambridge, MA: MIT Press.

Oberthür, S., and O. S. Stokke, eds. 2011. *Managing Institutional Complexity: Regime Interplay and Global Environmental Change*. Cambridge, MA: MIT Press.

Olson, M. 1965. *The Logic of Collective Action*. Cambridge, MA: Harvard University Press.

Open Working Group on Sustainable Development Goals. 2014. Outcome Document available at: http://sustainabledevelopment.un.org/focussdgs.html/.

Orsini, A., J.-F. Morin, and O. R. Young. 2013. "Regime Complexes: A Buzz, a Boom, or a Boost for Global Governance." *Global Governance* 19:27–39.

Ostrom, E. 1990. *Governing the Commons: The Evolution of Institutions for Collective Action*. Cambridge: Cambridge University Press.

Ostrom, E., et al., eds. 2002. *The Drama of the Commons*. Washington, DC: National Academies Press.

Ostrom, E., et al. 2007. "A Diagnostic Approach to Going Beyond Panaceas." *Proceedings of the National Academy of Sciences of the United States of America* 104:15181–15187.

Paddock, L., et al., eds. 2012. *Compliance and Enforcement in International Law*. Cheltenham, UK: Edward Elgar.

Paris Agreement 2015. Agreement adopted at UNFCCC COP 21. FCCC/CP/2015/L.9/Rev.1.

Park, J., K. Conca, and M. Finger. 2008. *The Crisis of Global Environmental Politics: Towards a New Political Economy of Sustainability*. London: Routledge.

Parson, E. A. 2003. *Protecting the Ozone Layer: Science and Strategy*. New York: Oxford University Press.

Passmore, J. 1994. *Man's Responsibility for Nature: Ecological Problems and Western Traditions*. New York: Charles Scribner's Sons.

Perrow, C. 1984. *Normal Accidents—Living with High-Risk Technologies*. Princeton: Princeton University Press.

Piketty, T. 2014. *Capital in the 21st Century*. Cambridge, MA: Harvard University Press.

Ponting, C. 2007. *New Green History of the World: The Environment and the Collapse of Great Civilizations*. London: Vintage.

Poteete, A. R., M. A. Janssen, and E. Ostrom. 2010. *Working Together: Collective Action, the Commons, and Multiple Methods in Practice*. Princeton: Princeton University Press.

Putnam, R. 1988. "Diplomacy and Domestic Politics: The Logic of Two-Level Games." *International Organization* 42:427–460.

Ragin, C. C. 1987. *The Comparative Method*. Berkeley: University of California Press.

Ragin, C. C. 2000. *Fuzzy-Set Social Science*. Chicago: University of Chicago Press.

Rapoport, A. 1960. *Fights, Games, and Debates*. Ann Arbor: University of Michigan Press.

Raskin, P., T. Banuri, G. Gallopin, P. Gutman, A. Hammond, R. Kates, and R. Swart. 2002. *Great Transition: The Promise and Lure of the Times Ahead*. Boston: Stockholm Environment Institute.

Raustiala, K., and A. M. Slaughter. 2002. "International Law, International Relations and Compliance," 538–558. In W. Carlnaes, T. Risse, and B. A. Simmons, eds., *Handbook of International Relations*. London: Sage Publications.

Raustiala, K., and D. G. Victor. 2004. "The Regime Complex for Plant Genetic Resources." *International Organization* 55:277–309.

Rawls, J. 1971. *A Theory of Justice*. Cambridge, MA: Harvard University Press.

Reich, R. 2015. *Saving Capitalism: For the Many, Not the Few*. New York: Vintage.

Riker, W. 1962. *The Theory of Political Coalitions*. New Haven: Yale University Press.

Ringquist, E. J., and T. Kostadinova. 2005. "Assessing the Effectiveness of International Environmental Agreements: The Case of the 1985 Helsinki Protocol." *American Journal of Political Science* 49:86–102.

Rio Declaration. 1992. Outcome document from the UN Conference on Environment and Development. UN Doc. A/CONF.151/5/Rev. 1.

Risen, J. 2014. *Pay Any Price: Greed, Power, and Endless War*. New York: Houghton Mifflin Harcourt.

Rockström, J., et al. 2009. "A Safe Operating Space for Humanity." *Nature* 461:472–475.

Rockström, J., and M. Klum. 2015. *Big World, Small Planet: Abundance within Planetary Boundaries*. New Haven: Yale University Press.

Rosenau, J. N., and E.-O. Czempiel, eds. 1992. *Governance without Government: Order and Change in World Politics*. Cambridge: Cambridge University Press.

Rothenberg, J. 1993. "Economic Perspective on Time Comparisons: Alternative Approaches to Time Comparisons," 355–397. In N. Choucri, ed., *Global Accord: Environmental Challenges and International Responses*. Cambridge, MA: MIT Press.

Rudel, T. K. 2005. *Tropical Forests: Regional Paths of Destruction and Regeneration in the Late 20th Century*. New York: W.W. Norton.

Ruggie, J. G. 1982. "International Regimes, Transactions, and Change: Embedded Liberalism in the Postwar Economic Order." *International Organization* 36:379–415.

Ruggie, J. G. 1998. *Constructing the World Polity: Essays on International Institutionalism*. London: Routledge.

Sabin, P. 2014. *The Bet: Paul Ehrlich, Julian Simon, and our gamble over Earth's future*. New Haven: Yale University Press.

Sachs, J. 2015. *The Age of Sustainable Development*. New York: Columbia University Press.

Sand, P. H. 1994. "Trusts for the Earth: New Financial Mechanisms for International Environmental Protection," Josephine Onoh Memorial Lecture, University of Hull.

Sands, P. H. 1995. "International Law in the Field of Sustainable Development: Emerging Legal Principles," 53–66. In W. Lang, ed., *Sustainable Development and International Law*. London: Graham and Trotman/Martinus Nijhoof.

Sand, P. H. 1995. "Trusts for the Earth: International Financial Mechanisms for Sustainable Development," 167–184. In W. Lang, ed., *Sustainable Development and International Law*. London: Graham and Trotman/Martinus Nijhoff.

Scheffer, M. 2009. *Critical Transitions in Nature and Society*. Princeton: Princeton University Press.

Schelling, T. C. 1960. *The Strategy of Conflict*. Cambridge, MA: Harvard University Press.

Schelling, T. C. 1966. *Arms and Influence*. New Haven: Yale University Press.

Schelling, T. C. 1978. *Micromotives and Macrobehavior*. New Haven: W.W. Norton.

Schellnhuber, H.-J. 2009. "Tipping Elements in the Earth System." *Proceedings of the National Academy of Sciences of the United States of America* 106:20561–20563.

Schellnhuber, H.-J., and H. Held. 2002. "How Fragile is the Earth System?" 5–34. In J. C. Briden and T. E. Downing. eds., *Managing the Earth*. Oxford: Oxford University Press.

Schellnhuner, H.-J., et al., eds. 2004. *Earth System Analysis for Sustainability*. Cambridge, MA: MIT Press.

Shabecoff, P. 1996. *A New Name for Peace: International Environmentalism, Sustainable Development, and Democracy*. Hanover, NH: University Press of New England.

Shepherd, K., et al. 2015. "Development Goals Should Enable Decision-Making." *Nature* 323:132–134.

Sikora, R. I., and B. Barry, eds. 1978. *Obligations to Future Generations*. Philadelphia: Temple University Press.

Silver, N. 2012. *The Signal and the Noise: Why So Many Predictions Fail—but Some Don't*. New York: The Penguin Press.

Simon, J. 1981. *The Ultimate Resource*. Princeton: Princeton University Press.

Singer, P. 1977. *Animal Liberation: A New Ethic for Our Treatment of Animals*. New York: Avon Books.

Singh, E., and A. Saguirian. 1993. "The Svalbard Archipelago: The Role of Surrogate Negotiators," 22–55. In O. R. Young and G. Osherenko, eds., *Polar Politics: Creating International Environmental Regimes*. Ithaca, NY: Cornell University Press.

Skjaerseth, J. B., O. S. Stokke, and J. Wettestad. 2006. "Soft Law, Hard Law, and Effective Implementation of International Environmental Norms." *Global Environmental Politics* 6:104–120.

Social Learning Group. 2001. *Learning to Manage Global Environmental Risks*, 2 vols. Cambridge, MA: MIT Press.

Speth, J. G. 2004. *Red Sky at Morning: America and the Crisis of Global Government*. New York: Oxford University Press.

Steffen, W. 2011. "Climate Change: A Truly Complex and Diabolical Policy Problem," 21–37. In J. Dryzek et al., eds. *Oxford Handbook of Climate Change and Society*. Oxford: Oxford University Press.

Steffen, W., et al. 2004. *Global Change and the Earth System: A Planet under Pressure*. Heidelberg: Springer Verlag.

Steffen, W., P. J. Crutzen, and J. R. McNeil. 2007. "The Anthropocene: Are Humans Now Overwhelming the Great Forces of Nature." *Ambio* 36:614–621.

Steffen, W., J. Grinevald, P. Crutzen, and J. R. McNeill. 2011. "The Anthropocene: Conceptual and Historical Perspectives." *Philosophical Transactions of the Royal Society A* 369:842–867.

Steffen, W., et al. 2011. "The Anthropocene: From Global Change to Planetary Stewardship." *Ambio* 40:739–761.

Steffen, W., et al. 2015. *The Trajectory of the Anthropocene: The Great Acceleration*. The Anthropocene Review. doi:10.1177/2053019614564785.

Steffen, W., et al. 2015. "Planetary Boundaries: Guiding Human Development on a Changing Planet." *Science*. doi:10.1126/scence.1259855.

Stern, N. 2007. *The Economics of Climate Change*. Cambridge: Cambridge University Press.

Stern, N. 2009. *The Global Deal: Climate Change and the Creation of a New Era of Progress and Prosperity*. New York: Public Affairs.

Stern, P., O. R. Young, and D. Drukman, eds. 1991. *Global Environmental Change: Understanding the Human Dimensions*. Washington, DC: National Academies Press.

Stewart, D. O. 2008. *The Summer of 1787: The Men Who Invented the Constitution*. New York: Simon and Schuster.

Stewart, D. O. 2015. *Madison's Gift: Five Partnerships that Built America*. New York: Simon and Schuster.

Stockholm Declaration. 1972. Outcome document from the UN Conference on the Human Environment. UN Doc. A/CONF.48/14/Rev.1.

Stokes, D. 1997. *Pasteur's Quadrant: Basic Science and Technological Innovations*. Washington, DC: Brookings Institution.

Stokke, O. S. 2001. "The Interplay of International Regimes: Putting Effectiveness Theory to Work," Report No. 14. Lysaker, Norway: Nansen Institute.

Stokke, O. S. 2004. "Boolean Analysis, Mechanisms, and the Study of Regime Effectiveness," 87–120. In A. Underdal and O. R. Young, *Regime Consequences*. Dordrecht: Kluwer Academic Publishers.

Stokke, O. S. 2011. "Interplay Management: Niche Selection and Arctic Environmental Governance," 143–170. In S. Oberthür and O. S. Stokke, eds., *Institutional Interaction and Global Environmental Change*. Cambridge, MA: MIT Press.

Stokke, O. S. 2012. *Disaggregating International Regimes: A New Approach to Evaluation and Comparison*. Cambridge, MA: MIT Press.

Stone, D. P. 2015. *The Changing Arctic Environment: The Arctic Messenger*. New York: Cambridge University Press.

Strange, S. 1983. "*Cave! hic dragones*: A Critique of Regime Analysis," 337–354. In S. Krasner, ed. *International Regimes*. Ithaca, NY: Cornell University Press.

Tainter, J. 1988. *The Collapse of Complex Societies*. Cambridge: Cambridge University Press.

Thaler, R. H. 2015. *Misbehaving: The Making of Behavioral Economics*. New York: W.W. Norton.

Thaler, R. H., and C. R. Sunstein. 2008. *Nudge: Improving Decisions About Health, Wealth, and Happiness*. New Haven: Yale University Press.

Tuchman, B. 1962. *The Guns of August*. New York: Macmillan.

Tolba, M. K. 1998. *Global Environmental Diplomacy: Negotiating Environmental Agreements for the World, 1973–1992*. Cambridge, MA: MIT Press.

Tomz, M. 2007. *Reputation in International Cooperation*. Princeton: Princeton University Press.

Transparency International (ongoing). Transparency International: The Global Coalition against Corruption. Information available at: https://www.transparency.org.

Turner, B. L., et al. 1990. *The Earth as Transformed by Human Action: Global and Regional Changes in the Biosphere over the Past 300 Years*. Cambridge: Cambridge University Press.

Tversky, A., and D. Kahneman. 1974. "Judgment under Uncertainty: Heuristics and Biases." *Science* 185:1124–1131.

Underdal, A. 2002. "One Question: Two Answers," 3–45. In E. L. Miles et al., *International Regime Effectiveness*. Cambridge, MA: MIT Press.

Underdal, A. 2008. "Determining the Causal Significance of Institutions: Accomplishments and Challenges," 49–78. In O. R. Young, L. King, and H. Schroeder, eds., *Institutions and Environmental Change*. Cambridge, MA: MIT Press.

Underdal, A. 2010. "Complexity and Challenges of Long-term Environmental Governance." *Global Environmental Change* 20:386–393.

Underdal, A., and O. R. Young, eds. 2004. *Regime Consequences: Methodological Challenges and Research Strategies*. Dordrecht: Kluwer Academic Publishers.

UNFCCC. 1992. United Nations Framework Convention on Climate Change. Text available at: www.unfccc.int.

UN General Assembly (UNGA). 2000. "United Nations Millennium Declaration," A/RES/55/2. http://www.un.org/millennium/declaration/areS552E.HTM.

UN General Assembly (UNGA). 2015. "Transforming Our World: The 2030 Agenda for Sustainable Development," A/RES/70/1.

United Nations. 2012. *The Future We Want*. Outcome document from UN Conference on Sustainable Development (Rio+20), UNGA Res. 66/288, July 27.

United Nations. 2015. *The Millennium Development Goals Report 2015*. UNDP_MDG_Report_2015.pdf.

United Nations Development Programme (UNDP). 2006. *Human Development Report 2006 – Beyond Scarcity: Power, poverty and the global water crisis*. New York: Palgrave Macmillan.

United Nations Development Programme (UNDP). 2007. *Human Development Report 2007–2008—Fighting Climate Change: Human solidarity in a divided world*. New York: Palgrave Macmillan.

United Nations Development Programme (UNDP). 2014. Discussion Paper: "Governance for Sustainable Development: Integrating Governance in the Post–2015 Development Framework," available at: www.undp.org/.

VanderZwaag, D. L. 2012. "The IMO and Arctic Marine Environmental Protection: Tangled Currents, Sea of Challenges," 99–128. In O. R. Young, J. D. Kim, and Y. H. Kim, eds., *The Arctic in World Affairs: A North Pacific Dialogue on Arctic Marine Issues*. Seoul and Honolulu, KMI and EWC.

Vatn, A. 2012. "Environmental Governance: The Aspect of Coordination," 31–53. In E. Brousseau et al., eds., *Global Environmental Commons*. Oxford: Oxford University Press.

Velders, G. J., et al. 2007. "The Importance of the Montreal Protocol in Protecting Climate." *Proceedings of the National Academy of Sciences of the United States of America* 104:4814–4819.

Victor, D. G. 2011. *Global Warming Gridlock: Creating Effective Strategies for Protecting the Planet*. Cambridge: Cambridge University Press.

Victor, D. G., K. Raustiala, and E. B. Skolnikoff, eds. 1998. *The Implementation and Effectiveness of International Environmental Commitments*. Cambridge, MA: MIT Press.

Vitousek, P., et al. 1997. "Human Domination of the Earth's Ecosystems." *Science* 277:494–499.

Vogel, D. 2006. *The Market for Virtue: The Potential and Limits of Corporate Social Responsibility*. Washington, DC: Brookings.

von Moltke, K. 1997. "Institutional Interactions: The Structure of Regimes for Trade and the Environment," 247–272. In O. Young, ed., *Global Governance*. Cambridge, MA: MIT Press.

Walker, B., and D. Salt. 2006. *Resilience Thinking: Sustaining Ecosystems and People in a Changing World*. Washington, D.C.: Island Press.

Wapner, P. 1996. *Environmental Activism in World Civic Politics*. Albany: State University of New York Press.

Wapner, P. 1997. "Governance in Global Civil Society," 65–84. In O. R. Young, ed., *Global Governance*. Cambridge, MA: MIT Press.

Wassmann, P., and T. M. Lenton. 2012. "Arctic Tipping Points in an Earth System Perspective." *Ambio* 41:1–9.

Waters, C. N., et al. 2016. "The Anthropocene is functionally and stratigraphically distinct from the Holocene." *Science* 8 (January). doi:10.1126/science.aad2622.

Wettestad, J. 2002. *Clearing the Air—European Advances in Tackling Acid Rain and Atmospheric Pollution*. Aldershot, UK: Ashgate.

Wiener, N. 1948. *Cybernetics or Control and Communication in the Animal and the Machine*. Cambridge, MA: MIT Press.

Wilson, E. O. 2007. *The Future of Life*. New York: Vintage.

Wilson, J. 2006. "Matching Social and Ecological Systems in Complex Ocean Fisheries." *Ecology and Society* 11:9.

Wolf, C. 1988. *Markets or Governments: Choosing between Imperfect Alternatives*. Cambridge, MA: MIT Press.

World Charter for Nature. 1982. UNGA A/Res/37/7.

World Commission on Environment and Development. 1987. *Our Common Future*. New York: Oxford University Press.

Young, O. R. 1979. *Compliance and Public Authority: A Theory with International Applications*. Baltimore: Johns Hopkins University Press.

Young, O. R. 1982a. *Resource Regimes: Natural Resources and Social Institutions*. Berkeley: University of California Press.

Young, O. R. 1982b. "Regime Dynamics: The Rise and Fall of International Regimes." *International Organization* 36:277–297.

Young, O. R. 1989a. *International Cooperation: Building Regimes for Natural Resources and the Environment*. Ithaca,, NY: Cornell University Press.

Young, O. R. 1989b. "The Politics of International Regime Formation: Managing Natural Resources and the Environment." *International Organization* 43:349–375.

Young, O. R. 1991. "Political Leadership and Regime Formation: On the Development of Institutions in International Society." *International Organization* 45:291–309.

Young, O. R. 1994a. *International Governance: Protecting the Environment in a Stateless Society*. Ithaca, NY: Cornell University Press.

Young, O. R. 1994b. "The Problem of Scale in Human/Environment Relations." *Journal of Theoretical Politics* 6:429–447.

Young, O. R. 1996a. "Institutional Linkages in International Society: Polar Perspectives." *Global Governance* 2:1–24.

Young, O. R. 1998. *Creating Regimes: Arctic Accords and International Governance*. Ithaca, NY: Cornell University Press.

Young, O. R. 1999. *Governance in World Affairs*. Ithaca, NY: Cornell University Press.

Young, O. R. 2001a. "The Behavioral Effects of Environmental Regimes: Collective-Action vs. Social-Practice Models." *International Environmental Agreement: Politics, Law and Economics* 1:9–29.

Young, O. R. 2001b. "Environmental Ethics in International Society," 161–193. In J-M. Coicaud and D. Warner, eds., *Ethics and International Affairs*. Tokyo: UNU Press.

Young, O. R. 2002a. *The Institutional Dimensions of Environmental Change: Fit, Interplay, and Scale*. Cambridge, MA: MIT Press.

Young, O. R. 2002b. "Are Institutions Intervening Variables or Basic Causal Forces? Causal Clusters vs. Causal Chains in International Society," 176–191. In M. Brecher and F. Harvey, eds., *Millennium Reflections on International Studies*. Ann Arbor: University of Michigan Press.

Young, O. R. 2003. "Determining Regime Effectiveness: A Comment on the Oslo-Potsdam Solution." *Global Environmental Politics* 3:97–104.

Young, O. R. 2005a. "Why Is There No Unified Theory of Environmental Governance?" 170–184. In P. Dauvergne, ed., *Handbook of Global Environmental Politics*. Cheltenham, UK: Edward Elgar.

Young, O. R. 2005b. "Governing the Bering Sea Region," 194–209. In S. Ebbin, A. H. Hoel, and A. Sydnes, eds., *The Exclusive Economic Zone and Governance Institutions for Living Marine Resources*. Dordrecht: Springer.

Young, O. R. 2008. "Building Regimes for Socioecological Systems: Institutional Diagnostics," 115–144. In O. Young, L. King, and H. Schroeder, eds., *Institutions and Environmental Change*. Cambridge, MA: MIT Press.

Young, O. R. 2008a. "The Architecture of Global Environmental Politics: Bringing Science to Bear on Policy." *Global Environmental Politics* 8:14–32.

Young, O. R. 2010. *Institutional Dynamics: Emergent Patterns in International Environmental Governance*. Cambridge, MA: MIT Press.

Young, O. R. 2011a. "The Effectiveness of International Environmental Regimes: Existing Knowledge, Cutting-edge Themes, and Research Strategies." *Proceedings of the National Academy of Sciences of the United States of America* 108:19853–19860.

Young, O. R. 2012a. "Navigating the Sustainability Transition," 80–101. In E. Brousseau et al., eds., *Global Environmental Commons*. Oxford: Oxford University Press.

Young, O. R. 2012b. "Arctic Tipping Points: Governance in Turbulent Times." *Ambio* 41:75–84.

Young, O. R. 2013a. "Sugaring Off: Enduring Insights from Long-Term Research on Environmental Governance." *International Environmental Agreement: Politics, Law and Economics* 13:87–105.

Young, O. R. 2013b. *On Environmental Governance: Sustainability, Efficiency, and Equity*. Boulder, CO: Paradigm Publishers.

Young, O. R. 2014a. "Does Fairness Matter in International Environmental Governance? Creating an Effective and Equitable Climate Regime," 16–28. In T. Cherry et al., eds., *Towards a New Climate Agreement*. London: Routledge.

Young, O. R. 2014b. "Navigating the Arctic/Non-Arctic Interface: Avenues of Engagement," 225–250. In Oran R. Young, Jong-Deog Kim, and Yoon Hyung Kim. eds., *The Arctic in World Affairs: A North Pacific Dialogue on International Cooperation in a Changing Arctic*. Seoul and Honolulu: KMI and EWC.

Young, O. R. 2016. "The Co-production of Knowledge about International Governance: Life on the Science/Policy Interface." In S. Nicholson and S. Jinnah, eds., *New Earth Politics*, 75–95. Cambridge, MA: MIT Press.

Young, O. R. 2016. "International Relations in the Anthropocene: The Twilight of the Westphalian Order." In K. Booth and T. Erskine, eds. *International Relations Theory Today*, 231–251. Cambridge: Polity.

Young, O. R. Forthcoming. "Goal-Setting and Rule-Making as Strategies for Earth System Governance." In N. Kanie and F. Biermann, eds, *Governing through Goals*. Cambridge, MA: MIT Press.

Young, O. R., and G. Osherenko, eds. 1993. *Polar Politics: Creating International Environmental Regimes*. Ithaca, NY: Cornell University Press.

Young, O. R., ed. 1999b. *The Effectiveness of International Environmental Regimes: Causal Connections and Behavioral Mechanisms.* Cambridge, MA: MIT Press.

Young, O. R., A. Agrawal, L. A. King, P. H. Sand, A. Underdal, and M. Wasson. 1999/2005. *Institutional Dimensions of Global Environmental Change (IDGEC) Science Plan. Reports Nos. 9 and 16.* Bonn: IHDP.

Young, O. R., et al. 2006a. "A Portfolio Approach to Analyzing Complex Human-Environment Interactions: Institutions and Land Use." [published online.] *Ecology and Society* 11:31.

Young, O. R., et al. 2006b. "The Globalization of Socio-Ecological Systems: An agenda for Scientific Research." *Global Environmental Change* 16:304–316.

Young, O. R., et al. 2015. "Institutionalized Governance Processes: Comparing Environmental Problem Solving in China and the United States." *Global Environmental Change* 31:163–173.

Young, O. R., L. A. King, and H. Schroeder, eds. 2008. *Institutions and Environmental Change: Principal Findings, Applications, and Research Frontiers.* Cambridge, MA: MIT Press.

Young, O. R., and W. Steffen. 2009. "The Earth System: Sustaining Planetary Life-Support Systems," 319–337. In F. S. Chapin, G. Kofinas, and C. Folke, eds., *Principles of Ecosystem Stewardship.* New York: Springer.

Zaelke, D., et al., eds. 2005. *Making Law Work: Environmental Compliance and Sustainable Development,* 2 vols. London: Cameron May.

Zaelke, D., et al. 2016. "Primer on HFC," IGSD Working Paper, April.

Zelli, F., and H. van Asselt. 2013. "Introduction, The Institutional Fragmentation of Global Environmental Governance: Causes, Consequences, and Responses." *Global Environmental Politics* 13:1–13.

Index